Crimes and
Mathdemeanors

Crimes and Mathdemeanors

Leith Hathout

Illustrated by Karl H. Hofmann

A K Peters, Ltd.
Wellesley, Massachusetts

CRC Press
Taylor & Francis Group
6000 Broken Sound Parkway NW, Suite 300
Boca Raton, FL 33487-2742

© 2007 by Taylor & Francis Group, LLC
CRC Press is an imprint of Taylor & Francis Group, an informa business

No claim to original U.S. Government works
Printed in the United States of America on acid-free paper
Cover illustration by Karl H. Hofmann.

International Standard Book Number: 978-1-56881-260-1 (Softcover)

**Library of Congress Cataloging-in-Publication Data
is available**

**Visit the Taylor & Francis Web site at
http://www.taylorandfrancis.com**

**and the CRC Press Web site at
http://www.crcpress.com**

Table of Contents

Preface	vii
Acknowledgments	xi
A Mystery on Sycamore Lane	1
The Watermelon Swindle	21
An Adventure at the Grand Canyon	27
Basketball Intrigue	39
The Moon Rock	55
A Theft at Dubov Industries	73
Murder at The Gambit	83
A Day at the Racetrack	95
Bowling Average	109
Caught on Film	119
A Mishap at Shankar Chemicals	133
Almost Expelled	147
The Urban Jungle	161
A Snowy Morning on Oak Street	181
Conclusion	191
About the Problems	193

Preface

> The great book of nature lies ever open before our eyes
> and the true philosophy is written in it ... But we can-
> not read it unless we have first learned the language
> and the characters in which it is written ... It is writ-
> ten in mathematical language....
>
> —Galileo Galilei

Quotes like the one above convey the immense value of
mathematics as an indispensable tool in understanding
the laws that govern our physical world. However, they also
make math sound serious and daunting and intimidating.
I like math not because it is valuable, but because it is fun
and because it is beautiful.

I love the "Aha" feeling as a flash of insight suddenly
illuminates the solution to some problems. I also love the
sense of exhausted victory as a relentless mental effort
provides the solution to others. Mostly, though, I love how
math is constantly full of surprise, sometimes completely
defying my intuition and overturning my common sense.

When I was in grade school, my favorite books to read
were the *Encyclopedia Brown* series by Donald J. Sobol. I
was amazed at how Encyclopedia Brown, using a mixture
of knowledge and logic, was able to solve mysteries that

vexed others. As I got older, and became involved in mathematics, I began to envision a kid detective like Encylopedia who would solve mysteries based not just on logic, but also on serious mathematics. I started playing with a few story ideas to see if such a concept would actually work. The book that you have before you is my attempt to render such a character and to convey the sense of mystery and joy that I find in mathematics. Interestingly, during the process of writing, a popular television series, *Numb3rs*, debuted, giving me hope that the idea of a mathematical detective may indeed be appealing to a broad audience.

My intended audience is my friends and others like them—young people who like math, but probably not enough to just sit and read mathematics texts or work through problem books. Yet, they may be amenable to thinking mathematically if it is within the context of solving a mystery or a puzzle—if a challenge is laid before them to match wits with the protagonist of the stories, Ravi. Therefore, the mathematics used in this book is mostly of the high-school level, with the hope that no reader will feel that the mathematics is beyond his or her understanding.

My use of the name "Ravi" for the young mathematical detective who solves these mysteries probably also deserves some comment. He is named after Ravi Vakil, a faculty member of the mathematics department at Stanford University. As I got more "serious" about math, one of my favorite books was Dr. Vakil's *A Mathematical Mosaic: Patterns & Problem Solving* (Brendan Kelly Publishing, Burlington, Ontario, 1997). It has just the right flavor of fun and rigor and has very interesting profiles of young mathematicians that Ravi Vakil had come to know. Thus, I decided to name my hero after Dr. Vakil. I must clarify that

Dr. Vakil does not know me, and I never consulted with him before naming my detective. I sincerely hope that he does not mind my choice.

The problems around which the detective stories in this book are built come from many sources, which I have tried to reference in a separate index. I highly recommend all of these books to anyone with some interest in mathematics. Unfortunately, over a few years of reading about mathematics and solving math problems, I sometimes could remember a problem but not where I had seen it. Whenever I could find the source of the problem at the heart of a story, I have cited it. Where I could not, I ask the forgiveness of those authors whom I did not credit. Another complicating factor is that problems are often found in many books, especially when they have become popular and part of the "lore" of mathematics. This makes it difficult to assign original credit to any one person for a problem. In any case, I make no pretense of developing most of these problems myself— only a few are my own. I have, however, adapted them as needed and placed them within the context of interesting and novel stories. Also, I have tried to present the solutions in an original fashion and one suited to my intended audience.

I hope you, the reader, find in this book some of the pleasure that I had in writing it. If it does for some what *A Mathematical Mosaic* did for me—igniting a spark of enthusiasm about mathematics—I will be both humbled and proud.

Acknowledgments

I realize that it is unusual for someone my age to write a book, let alone a mathematics book. It would not have been possible without the encouragement of my parents, as well as the tremendous support of my wonderful publisher, Klaus Peters. I am very grateful to Dr. Katherine Socha for reviewing the entire manuscript in detail and presenting many helpful suggestions as well as pointing out and correcting errors as they occurred. Also, I sincerely thank Dr. Karl Hofmann for also reviewing the stories and providing many helpful comments, and most especially for his wonderful and vibrant illustrations that truly brought the stories to life. To Ms. Charlotte Henderson, I owe an inestimable debt of gratitude. She shepherded the manuscript to production, with meticulous attention to detail. She reviewed, edited, and clarified where needed, and she also redrew all of the figures. The book is prettier, sharper, crisper, and better because of her extensive efforts.

Finally, I would like to especially thank my father. When he read my first stories, he was the one who suggested that the idea of a mathematical Encyclopedia Brown-like detective had the makings of a book. He is a physician who, at the time, was engrossed in writing a medical textbook. He encouraged me to pursue the idea of a book, promising that we would find the discipline to work together, I on my book and he on his. He kept his promise.

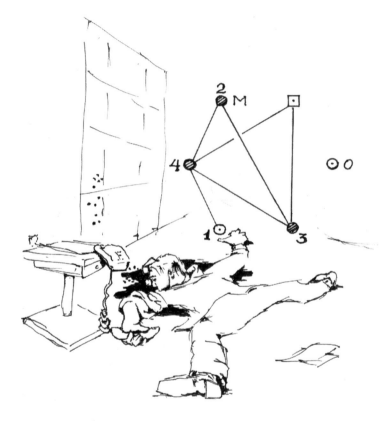

A Mystery on Sycamore Lane

Ravi strode onto the large, neatly manicured lawn of 423
Sycamore Lane. Two police cars with flashing lights stood
in the driveway of the large white house with tall columns
at its entrance. The policemen nodded to Ravi as he walked
by. He was a slender fourteen-year-old boy with curly brown
locks and large, inquisitive brown eyes. While it was not
common for the police to let a teenager stroll into the scene
of a homicide investigation, Ravi was no common teenager.
He was, by all accounts, a genius, though he was far too
modest to think of himself in that way. Rather, he thought of
himself as someone who loved puzzles, and loved thinking
through problems, especially if they had a mathematical
flavor. His uncanny ability to do just that had earned him
the trust of Chief Dobson, head of the Criminal Investiga-
tions Unit of the Chicago Police Department. Chief Dobson
had come to know Ravi through Andy, his son and Ravi's
classmate in tenth grade. While Ravi was at the Dobsons'
house for dinner one evening, the Chief was recounting the
details of an especially difficult case. Upon hearing the case,
Ravi was able to solve it with just a few minutes of thinking. Since then, Chief Dobson made it a habit to consult
Ravi when faced with a case that was especially puzzling.

Ravi quickly bounded up the four steps that led to the half-open large carved wooden door of the residence and walked into the marble entry.

"Hello Ravi. Good to see you," said Chief Dobson.

"Hello Chief," replied Ravi. "What have we got?"

Chief Dobson motioned Ravi to follow him into the living room.

"This is Dr. and Mrs. Arden. Unfortunately, a murder was committed in their house today," said Chief Dobson.

"I'm sorry to hear that," responded Ravi. "How can I help, Chief?"

"Well, the victim was a man named Dr. Rosmoyne, a friend of the Ardens. He was killed in the study with a single shot to the back of the head early this afternoon. Apparently, he was on the phone with his back to the door, since the receiver was dangling from its cord next to him when he was found."

"Who found him?" asked Ravi.

"I did," answered Dr. Arden. Ravi turned to face Dr. and Mrs. Arden, who were sitting clasping hands on a small sofa.

"Please tell me what happened," requested Ravi.

Dr. Arden began to recount the details of the barbecue lunch which had ended so tragically: "We invited some friends over for Sunday brunch. We invited Dan Rosmoyne, the Wentworths and the Finnegans. Our guests showed up between 11:00 and 11:30. We hung out between the backyard patio and the upstairs den. We just ate and talked and watched the basketball game, you know, just a typical lazy Sunday. I was out by the pool most of the time barbecuing, and Stacey (he looked at his wife) was back and forth between the kitchen and the den. Artie Wentworth and I were talking politics while I barbecued. The others were upstairs in the den watching the game. I didn't see Dan for most of the party. He didn't like the sun much. Anyway, I thought he had left—some sort of emergency at the hospital, I thought."

"Did anyone see him leave?" asked Ravi.

"No, not really," answered Dr. Arden. Mrs. Arden also meekly shook her head no.

"Why did you think there was an emergency at the hospital, Dr. Arden?" asked Ravi.

"I think Bob Finnegan mentioned something when he came out to check how I was doing with the barbecue. He likes his steak extra well-done. I guess Dan asked him where our phone was because he needed to make a call, and Bob told him that the phone was downstairs in the study. Bob asked if that was okay, and I told him that it was. Then we all got caught up in the basketball game—the Bulls went into double overtime. When I noticed he wasn't there, I assumed he had run out to the hospital. That's the life of an obstetrician, you know. Babies just can't wait."

"So he was your colleague at the hospital, Dr. Arden?" inquired Ravi.

"No, no. I'm a professor at the university; I have a PhD in sociology."

"Then how did you know the victim, Dr. Arden?"

There was an awkward silence. Dr. and Mrs. Arden looked at each other. Dr. Arden took a deep breath and began, "He was—uh—our doctor. We were having trouble conceiving a baby, and we went to him. We've been sort of friends since, and we see each other from time to time. We were not very close or anything, but he was friends with our other guests, and his wife was out of town, so we invited him to join us."

"How did you find him, Dr. Arden?" asked Ravi, scratching his chin softly.

"A little while after the guests left, I walked into the study, and there he was lying face down on the ground in a pool of blood. It was horrible!"

Ravi continued his inquiry while Chief Dobson looked on, not wishing to disturb him, although many of these questions had already been asked and answered, "Was there anyone else in the house? A maid, perhaps, or children?"

"No," answered Dr. Arden, "Sunday is the maid's day off. And, as I mentioned, we don't have kids. The Finnegans left their kids at home, and the Wentworths don't have kids either."

"But Julie Wentworth is pregnant," interjected Mrs. Arden.

"Do you remember which of your guests left first, Mrs. Arden?" asked Ravi.

"They all left together," answered Dr. Arden.

"Are you sure?" Ravi inquired.

"Yes. Stacey and I walked them out to their cars in the driveway and we shook hands and they left. I'm usually a bit absent-minded, but this I remember quite distinctly, because just before they got into their cars, I happened to ask the Finnegans and the Wentworths, as well as Stacey, how many hands they had just shaken—it's for a project I'm working on about shifts in our social customs—and everyone gave me a different answer, which I thought was rather curious."

Ravi's eyebrows raised slightly and he said distractedly, looking off into the air, "Do you remember those answers?"

"No. That's where the absent-minded stuff comes in, you know."

"Yes, of course," said Ravi, turning to Stacey Arden, "Do you remember what they said, Mrs. Arden?"

"No, I don't," she answered.

"Do you remember how many hands you shook, then?"

"Yes. Four. I shook hands with all my guests," she answered after some hesitation.

"What else do we know, Chief?" asked Ravi, looking at Chief Dobson. "Have you talked with the Finnegans and the Wentworths?"

"Yes, of course. Some of my men are still at their houses. They gave pretty much the same story. Bob Finnegan says Dan Rosmoyne asked him about a phone, and Finnegan told him that the only phone he knew of was in the study, but there was also probably one in the Ardens' bedroom. Then, everyone got caught up in the game, and figured Dr. Rosmoyne had left."

"Did anyone hear a gunshot?" asked Ravi.

"No, apparently not. But they all say the TV was up loud and they were screaming at the players. Michael Jordan wasn't shooting too well but was hogging the ball," answered the Chief.

The Chief then cocked his head, indicating to Ravi that he wanted him to step out into the entry. The Chief followed him, and leaned in close, "We don't have a murder weapon. My men are over searching at the Finnegans and the Wentworths now. But we may have a motive. We found out that Bob Finnegan is an office administrator and that he used to work for Dr. Rosmoyne, but Rosmoyne had to let him go a couple of years ago when he switched his practice over to Mercy Hospital. Finnegan claims there were no hard feelings and that they were good friends. Mercy Hospital, apparently, has its own administrators, and that was the reason Rosmoyne let him go. Also, Rosmoyne used to date Julie Wentworth, before meeting his wife. But she says that was a long time ago, and that they mutually decided to stop seeing each other, and that they've remained good friends since."

"Interesting," mused Ravi. "Do we have anything else?"

"Yes," continued the Chief, still leaning low and half-whispering. "We ran paraffin tests for gunpowder residue on everyone, figuring there would be gunpowder on the hands of the shooter. Bob Finnegan tested positive. But I think we've got some contamination or something. Julie Wentworth and Stacey Arden also tested positive, but everyone else was negative."

"Was Dr. Rosmoyne wearing a pager, Chief?" asked Ravi.

"What?"

"A pager, a beeper, like doctors wear. I know it's Sunday, but did he have his beeper?"

The Chief looked vexed, "I don't know. We didn't notice it, but I'll call downtown to the medical examiner's office and have them check right now. I've got a man over there."

The Chief pulled out his cell phone and began to dial. Meanwhile, Ravi walked out into the cool air of the early evening, and strolled pensively along the lawn, thinking and looking at the lush green grass beneath his feet.

The Chief came running out. "Good thinking, Ravi. He did have a small beeper in his pocket. They checked if he had pages, and he had gotten a page at 12:49. The pager was on vibrate, so I guess no one heard him get beeped. When my officer dialed the number on the pager, it was the Mercy Hospital operator. So, like I said, good thinking, but it doesn't really get us anywhere. We're still stuck."

"No we're not, Chief," said Ravi as he walked toward the edge of the driveway to retrieve his bicycle. Chief Dobson knew that Ravi had to be home for Sunday dinner. "I think the case is solved," said Ravi with a little smile as he straddled his bike.

"Solved?!" asked Chief Dobson, wide-eyed with disbelief.

"Yes," said Ravi. He began to pedal out of the driveway and onto the street. He turned his head and said to Chief Dobson, "The murderer is ..."

Now, it is time for you to match wits with Ravi. Do you know who killed Dr. Rosmoyne?

Analysis

Later that evening, after Ravi had finished dinner with his parents, Chief Dobson drove to Ravi's house and asked him how he had solved the case. Ravi smiled and said, "Dr. Arden actually solved it for me, Chief. The case hinged on a little problem in logic—on how many hands Stacey Arden must have shaken."

Ravi continued, "Let's say you and your wife have a dinner party, and invite two other couples over. After dinner, you both walk them to the door, and people shake hands goodbye. Of course, a husband doesn't shake his own wife's hand, and vice versa. You happen to ask each of your guests as well as your wife how many hands they shook. Now, suppose each person gives you a different answer."

"Okay," said the Chief.

"Then how many hands did your wife shake?" asked Ravi.

"What?" answered the Chief, as if he did not comprehend the question.

"How many hands did your wife shake, Chief?" repeated Ravi.

"I don't know. You didn't tell me that. How can I possibly figure that out? I don't have enough information. Anyway, what does it matter? What does this have to do with our case?"

☙

Ravi was able to distill the details of the case into a single problem, the solution of which is also the solution to the case. Chief Dobson could not figure out the problem. If you can figure it out, you will solve the case as well!

Solution

To solve this problem, we need to "see through it." Here, it seems that there is not nearly enough information for Chief Dobson to determine how many hands his wife shook, or for us to determine how many hands Mrs. Arden shook. If we think carefully about the problem statement, though, a unique solution presents itself.

Since it seems that there is insufficient information, every piece of information we do have is critical. We are faced with a situation where Dr. Arden and his wife invite two other couples over. We know also that spouses don't shake hands. Now, two key facts need to be noticed:

1. The maximum number of hands which any person can shake is four. That is because there are six people total, made up of three couples. Person X does not shake hands with himself or his (or her) spouse, leaving a maximum of four possible handshakes.

2. Dr. Arden asked his wife and the two couples (five people total) about their handshakes and was quite definite that each person gave him a different answer. Since the maximum number of handshakes is four, and each of the five people shakes a different number of hands, the handshakes have to be distributed as zero, one, two, three, and four across Mrs. Arden and the two couples. We note that we have no information about how many hands Dr. Arden shook, since he did not ask himself any questions. Therefore, only the other five people need to be considered.

From these key facts, we can actually deduce another rule before moving forward. Consider the person who shakes

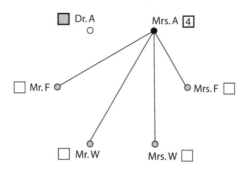

Figure 1. Mrs. A shakes four hands.

four hands. This person does not shake his own hand or the hand of his wife; therefore, he shakes hands with each of the four remaining people in the group. Then, this person's spouse must be the one who shakes zero hands, because everyone else has shaken the first person's hand. Therefore, as an additional principle, if someone has four handshakes then his (or her) spouse must have zero.

Now, we attempt to discover whether these facts provide us enough information to figure out how many hands Mrs. Arden actually shook. She claimed to have shaken four hands, and so we check that possibility first, using the diagram in Figure 1, where the individuals are identified by their initials.

If Mrs. Arden shook four hands, it is clear that none of the other guests could have shaken zero hands, since she shook hands with them all. Therefore, this possibility is eliminated.

Similarly, we can investigate whether Mrs. Arden could have shaken zero hands. In that case, one of the guests must have shaken four hands: let's say Mr. W. He cannot shake hands with Mrs. W, and so must have shaken hands with Mr. and Mrs. F and Dr. and Mrs. A. This contradicts the assumption that Mrs. A shook zero hands.

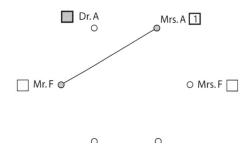

Figure 2. Mrs. A shakes one hand.

Now we can investigate the possibility that Mrs. Arden shook only one hand, again using a simple diagram (see Figure 2).

Since the situation is symmetric, it doesn't matter with whom Mrs. A shakes hands. Let's say it's Mr. F. She now cannot shake any more hands, because—by assumption—she shakes only one hand. Mr. F must be the person who shakes four hands, since no one else can shake Mrs. A's hand and thus has only three available handshakes. So, now we have the diagram in Figure 3.

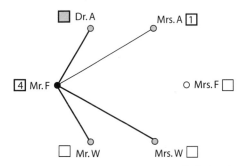

Figure 3. Mrs. A shakes one hand, and Mr. F shakes four hands.

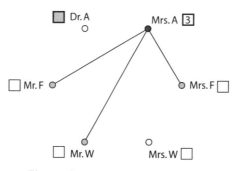

Figure 4. Mrs. A shakes three hands.

Since Mr. F shakes four hands, Mrs. F must be the person who shakes zero hands, by our additional principle. Now, Mr. and Mrs. W need to be assigned two and three handshakes. However, that is impossible. Let's look at Mr. W. (The situation for Mrs. W is symmetric.) He cannot shake hands with Mrs. W (his spouse) or with Mrs. F (she shakes zero hands) or with Mrs. A (she shakes only one hand by assumption and that is with Mr. F). Therefore, Mr. W can shake hands only with Dr. A and Mr. F, and it is impossible for him to shake three hands. Therefore, Mrs. A cannot have shaken just one hand.

Now, let us assume that Mrs. A shakes three hands, as in Figure 4. Therefore, she needs to shake hands with both members of one couple and with one member of the other couple. Let us assume that she shakes hand with Mr. and Mrs. F and with Mr. W. Therefore, Mrs. W has to have zero handshakes (everyone else, excluding Dr. A, has shaken Mrs. A's hand). It follows that Mr. W must be the one to shake four hands. (It cannot be Mr. and Mrs. F, because to get four handshakes, they would need to shake hands with Mrs. W, and we've already assigned her zero handshakes.)

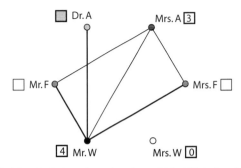

Figure 5. Mrs. A shakes three hands, and Mr. W shakes four hands.

Therefore, in addition to Mrs. A, Mr. W shakes hands with Dr. A, Mr. F and Mrs. F, as diagrammed in Figure 5.

Now, there is a problem: no one can be assigned just one handshake, since the two people who haven't been assigned a number—Mr. F and Mrs. F—already have two handshakes. Therefore, Mrs. A cannot have shaken three hands.

Now, we are left with only one option: Mrs. A shook two hands. If this is true, there are two possibilities: either she shook the hands of one couple, e.g., Mr. and Mrs. F, or she shook hands "across couples" with one partner from each, such as Mr. F and Mrs. W. Let us look at the first possibility, shown in Figure 6.

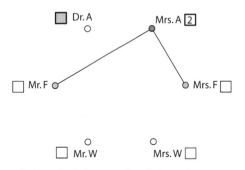

Figure 6. Mrs. A shakes two hands from the same couple.

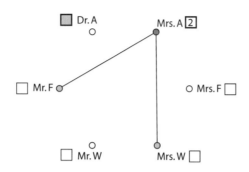

Figure 7. Mrs. A shakes two hands, one from each couple.

Now, one of the people who shook Mrs. A's hand has to be the person with four handshakes. (No one of the guests could shake four hands without shaking hers.) But we know that the person who shakes four hands has to be partnered with the person who shakes zero hands; in other words, if Mr. F shakes four hands, then Mrs. F has to shake zero. However, that is impossible, because by assumption Mrs. F has shaken hands with Mrs. A.

Now, we are down to one scenario, where Mrs. A shakes two hands, but not from the same couple. Since the situation is symmetric, let's say it's with Mr. F and Mrs. W, as in Figure 7.

Now, to complete this scenario, either Mr. F has four handshakes and Mrs. F has zero, or Mrs. W has four and Mr. W has zero. Again, since the situation is symmetric, let's assume that Mr. F has four handshakes and Mrs. F has zero, as in Figure 8.

Now, one and three handshakes have to go to Mr. and Mrs. W. Mrs. W is already shaking more than one hand, so Mr. W's one handshake is with Mr. F (as shown). Mrs. W's first two handshakes are with Mrs. A and Mr. F; since

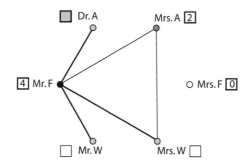

Figure 8. Mrs. A shakes two hands, and Mr. F shakes four hands.

Mrs. F shakes zero hands, Mrs. W's third handshake is with Dr. A. The complete situation is shown in Figure 9.

Now, it works, and there are no contradictions! The problem allows only this solution (or a symmetric variant). This is a surprising and wonderful result. We started with a situation wherein we thought that there was not enough information to solve the problem and progressed to a situation where we not only know that Mrs. Arden must

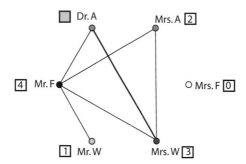

Figure 9. Situation with zero, one, two, three, and four handshakes.

have shaken precisely two hands, but that they must have belonged to members of different couples.

In this case, Mrs. Arden shook hands with Bob Finnegan and with Julie Wentworth. Mrs. Arden had shot Dr. Rosmoyne and gotten gunpowder residue on her right hand. She then contaminated the hands of Bob Finnegan and Julie Wentworth, who were her only two handshakes. It was her bad luck that their handshakes with her were their last. This allowed Ravi to solve the case.

After Ravi carefully explained the solution to Chief Dobson, he asked him to check Mrs. Arden's home and cell phone records for Sunday. Indeed, a call went out from Mrs. Arden's cell phone to Dr. Rosmoyne's pager at precisely 12:49. She had paged him to the Mercy Hospital number, to lure him into the study to answer his page. While everyone else was preoccupied with the game, she shot him. When faced with these facts, including that she had lied about the number of hands she shook, Mrs. Arden confessed. It turned out that she had always blamed Dr. Rosmoyne for failing to help her get pregnant. The Finnegans had been able to have kids. Then, when Mrs. Arden found out that Julie Wentworth had also gotten pregnant, she lost her senses and took her frustrations out on Dr. Rosmoyne.

Thus, Ravi put another case to rest. Now, it was time for bed, as Ravi had a math test Monday morning.

Extension

Thinking about this case actually allows us to prove a very interesting theorem:

> A host and his wife invite x couples to their home for dinner. Upon leaving, people shake hands, with the understanding that no one shakes hands with their own spouse. If the host asks his wife plus the other $2x$ people in attendance how many hands they shook, and everyone gives him a different answer, then his wife has shaken x hands.

The best way to prove this theorem is by mathematical induction. This method rests on two pillars:

1. Proving the theorem for a single base case, usually for some small number.
2. Showing that if the theorem is true for some case $x = n$, then it is also true for case $x = n + 1$.

These two sub-proofs, taken together allow us to show that it is true for all x.

We use the notation $P(x)$ to represent the number of handshakes for the host's wife if they have invited x couples and then everyone shakes hands according to the rules above. We want to prove that $P(x) = x$ for all possible x.

Our base case here is for $x = 2$, which we proved in the solution to the mystery, showing that $P(2) = 2$. This means that if two couples are invited, the wife of the host shakes two hands. Thus, we have shown that $P(x) = x$ for this base case of $x = 2$.

Now, if we show that "if the theorem is true for $x = n$, then it is true for $x = n + 1$ for any value of n," then the truth of $P(2) = 2$ implies that $P(3) = 3$, which in turn implies that $P(4) = 4$, etc., so that for any x, $P(x) = x$ as long as the conditions that no partners shake each other's hand and that everyone gives the host a different handshake number are satisfied.

Note that, under the conditions of the theorem, for x couples plus the host's wife, $2x + 1$ people are questioned. The maximum number of handshakes is $2x$. (Of the $2x + 2$ people at the party, including the host, the person shaking the maximum number of hands does not shake hands with himself or his spouse, leaving $2x$ possible handshakes.) The responses of the $2x + 1$ people questioned thus range from 0 to $2x$. Also remember that we have also shown that the person shaking $2x$ hands has to be paired with the person shaking 0 hands.

Now, let us say that $n + 1$ couples are invited. Then, there are $2 \times (n + 1) = 2n + 2$ maximum possible handshakes. There are $2n + 4$ people total at the party, including the host. He asks the $2n + 2$ guests plus his wife (for a total of $2n + 3$ people) how many hands they shook, and each person gives a different answer, ranging from 0 to $2n + 2$. (Remember that this totals to $2n + 3$ responses.) Let us consider person Mr. X at the party who shook the maximum number of hands (i.e., with everyone except himself and his wife). We know that his wife, Mrs. X, shook zero hands.

Now let's make Mr. and Mrs. X magically disappear before the handshakes. Thus, everyone at the party will shake precisely one less hand than they otherwise would have, because Mr. X shook hands with everyone, while Mrs. X shook hands with no one. Because Mr. and Mrs. X disappeared, this smaller party has n couples. We know, by assumption, that for a party of n couples, the wife of the host shakes n hands. If this is one less than she would have shaken had the Xs not disappeared, then if there were $n + 1$ couples, she would have shaken $n + 1$ hands; in other words,

$$P(n+1) - 1 = P(n),$$
$$\text{so } P(n+1) = n + 1 \text{ if } P(n) = n.$$

Thus, the proof is complete.

The Watermelon Swindle

"Slow down, Ravi, or you'll choke," pleaded his mother. Ravi was gulping down his dinner of meatloaf and peas. He was in a rush to get out the door.

"Tonight is the first meeting of the Computer Club for the year," explained Ravi through a mouthful of meatloaf. His mother shrugged and sighed.

Ravi's mother turned to her husband, who was reading at the dinner table.

"What does your day look like tomorrow, dear?"

"It should be pretty easy," he said, looking up from the legal memo he was reading. Ravi's father is a lawyer—actually a district attorney for Cook County in Illinois. "I just have one trial tomorrow, and it's an open and shut case. It should be over by early afternoon."

The father looked to Ravi and said, "This case should interest you, son. It involves your favorite fruit."

"Someone committed a crime with a watermelon, Dad?" asked Ravi, smiling.

"No, of course not," said his father. "It's a fraud case. A poor watermelon farmer is getting ripped off, and I'm suing the thieves for larceny."

"Interesting, Dad. How does a watermelon farmer get ripped off?" asked Ravi.

The father answered, "His name is Dimsdale. Mr. Dimsdale lives in Louisiana. He does business with Amex Grocers here in Chicago. They sell his watermelons to local grocery stores and give him the money. For this, he pays them 7 cents for every pound of watermelon that they sell, up front. Last month, he put a crop on a barge, which sailed up the Mississippi River. I have the manifest from the barge. He loaded two large cargo containers full of watermelons, with a total weight of 10,000 pounds, confirmed by the Louisiana Port Authority. With a wholesale price of watermelons at 83 cents per pound, he was expecting $8300.00 from Amex. He included a check for $700.00 for the transaction fee. They received the watermelons on August 12 and sold them to local grocers. When they paid Mr. Dimsdale, he received only $4140.04. Amex did not keep weight records of how much they sold to each grocer, or any receipts, but their manager claims that they turned over all the money they received to Mr. Dimsdale."

"Hmm," muttered Ravi.

"They had the nerve to say that the watermelons dehydrated in the sun on the barge up the Mississippi. Luckily for us, Amex had one watermelon from the crop which they had not sold, and Mr. Dimsdale had a watermelon from the same harvest which he had kept. I had Dimsdale's watermelon flown up here and had both watermelons analyzed at the crime lab," continued the father.

Ravi remained focused on his meatloaf, asking, "Did the watermelons actually dehydrate, Dad?"

"I have the lab reports right here. They dehydrated an insignificant amount. The watermelon from Amex is a Red Tiger watermelon, 98% water by weight. The watermelon from Dimsdale's original crop is a Red Tiger watermelon,

99% water by weight. Open and shut. I'm not going to let Amex's poor recordkeeping be used as an excuse to rip off a watermelon farmer."

Ravi finished his plate, stood up, and went towards the door. "Bye Mom. Bye Dad," he said. As he reached for the doorknob, he turned to his father and said, "By the way, Dad, I'd rethink that case if I were you."

Why did Ravi say that to his father?

Analysis

Watermelons are so named because they are 99% water by weight. Let's assume that 10,000 pounds of watermelons sat out in the sun for several hours and dehydrated, such that they became 98% water by weight. What do they weigh now?

Solution

Most people, when looking at this problem, solve it as follows: The watermelons were 99% water by weight, but now they are 98% water by weight, so they weigh 10,000 × (98/99) = 9898.99, for a loss of 101 pounds.

However, this is not the correct approach, because it ignores the fact that the percentages are referenced to an unknown weight which changes. Therefore, the analysis should be performed as follows.

If the watermelons are originally 99% water by weight, then they are 1% solid (seeds, sugars, rind, etc.). For an original weight of 10,000 pounds, the solid component thus weighs 0.01 × 10,000 = 100 pounds. After the watermelons dehydrate to be 98% water by weight, the solid component now makes up 2% of the new weight (w) of the watermelons, i.e.,

$$0.02w = 100 \text{ pounds, or } w = 5,000 \text{ pounds!}$$

Amazingly, the watermelons end up losing 50% of their weight.

Thus, Mr. Dimsdale, the watermelon farmer would indeed have been owed only 50% of the $8300.00 he expected, i.e., $4150.00. When we subtract from this $9.96 (12 pounds × 83 cents per pound) for the single unsold watermelon which happened to weigh 12 pounds, he should get $4140.04, which is precisely what he received!

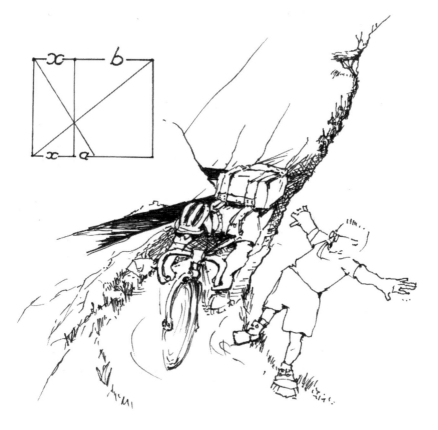

An Adventure at the Grand Canyon

Ravi strained his eyes against the thick darkness of the night. He was trying his best to see the small dirt roads that forked off the single paved road leading to Camp Ashenwood along the eastern-most edge of the Grand Canyon.

Ravi's family had been driving for hours and had decided to make the push into camp tonight, but the drive had taken much longer than anticipated. It was now nearly 2:00 a.m. and Ravi's father was fighting sleep—and apparently sleep was beginning to win.

"There, Dad, take a right," yelled Ravi.

The SUV veered to the right onto a dirt road and slowed down to a crawl. Even with the high beams on, the headlights seemed to barely make an impression in the enveloping darkness. Living in the city, it is difficult to remember how dark the wilderness gets at night when there is no moon.

The family's car continued on until Ravi's family spotted a dim light on the porch of a small cottage.

"There, Dad," pointed Ravi. "I think we're here."

His father barely nodded as he pulled the car in front of the cottage. Ravi's mother had long since fallen asleep in the back seat. Ravi and his father walked up to the cottage and knocked on the door. The door opened, and a man in a grey uniform stepped out into the dim light.

"Can I help you?" he asked.

Ravi's father proceeded to inform the ranger that the family had reserved a lot in which to camp. The father also gave the ranger the names of each of the family members, informing him that they intended to go on a hike tomorrow. Anyone who wanted to hike into the Grand Canyon had to give their names to the ranger the night before. This is so that if they get lost, a search and rescue mission could be launched. During the off-season, a ranger was on duty in the camp only during the night. The park service, although strapped for money, thought this was necessary for the safety of the campers. At this time of the year, the camp was nearly deserted. In fact, the camp only had a group of bicyclists who were there for training. They came to hike and ride the trail through the Grand Canyon that connected Camp Ashenwood on the north side with Camp Bramblewood on the south side.

Next morning, Ravi awoke with the sun shining on his face. Ravi knew that he had slept a long time and looked at his watch. "11:20. I don't remember the last time I slept this late," he thought to himself.

He woke his sleeping parents. He was anxious to start on their hike. However, it was clear that they had slept in far too long to take a serious hike today. They would have to be content with just seeing the canyon and exploring the first part of the trail. Ravi's parents insisted on having breakfast—now really lunch—before setting out on the trail. That meant setting up the portable stove (and a longer delay).

"Well, at least we'll get to see the bald eagle," said Ravi.

A few hundred yards into the trail, there was supposed to be a bald eagle's nest. Ravi had read about it in his hiker's guide. This was an especially exciting time because the eagle had laid eggs which should be close to hatching. It was a great opportunity to see such a rare species, especially at a time when there were eggs in the nest.

It was now 2:15 p.m., and Ravi and his parents finally set out on the trail heading from Camp Ashenwood to Camp Bramblewood. Ravi was trying to follow the map to the bald eagle's nest. It was inside an alcove along a craggy rock face adjacent to the trail. As Ravi turned a bend in the trail, he immediately jumped back. "Careful!" he said, as a man on a mountain bike almost hit him and then sped by. Ravi wondered how the man could bike so fast with such a large pack on his back.

"Excuse me," yelled Ravi's father after the man, but he did not respond.

"I wanted to ask him about directions to the nest," said the father, disappointed.

The family walked on. A few hundred feet down the trail, Ravi found the rock face, with the nest visible within an alcove about 30 feet up. There was no eagle in sight, however.

"What a shame," said Ravi's mother, "the eagle has flown away."

"No, Mom, that's impossible," said Ravi. "The eagle would not leave its eggs at such a critical time."

Ravi then started to climb the rock face, carefully finding his footing along the jagged rocks.

"Ravi, come down," pleaded his mother.

Within a few minutes, though, Ravi was up at the nest and he peered into it.

"The eagle has been stolen!" yelled Ravi. He carefully climbed down the rock face. "The eggs are gone too!"

That evening, when a ranger came on duty, Ravi and his parents reported the theft. The ranger was quite alarmed and immediately called the National Parks Service. The next morning, two federal agents had arrived at the camp site, having flown overnight from Washington, D.C. The agents had assembled Ravi and his family, Ranger Stinson, who was the ranger on duty the night Ravi's family drove into camp, and John Evers and Ward Thompson, the only other people who were on the trail the day the eagle was stolen, according to the log book in the rangers' cabins at Camp Ashenwood and Camp Bramblewood.

The agents asked in detail how and when Ravi had discovered the empty nest. They surmised that the eagle must have been stolen sometime after sunrise as no one could climb the rock face in the dark.

"What were you doing on the trail, Mr. Evers?" asked one of the agents.

"I am training for an upcoming race. We all are," he said, referring to the bicyclists who were at both camps. "We ride the trail up and down the canyon. I started from Bramblewood at sunrise and rode the trail into Ashenwood."

"What time did you get into Ashenwood?" asked the agent.

"I arrived at 2:30 p.m.," replied Evers.

"That's right," said Ravi. "He passed us on the trail at about 2:25: he almost hit me."

"I tried to flag you down, but you seemed in an awful hurry," said Ravi's father to Evers, recalling the big pack that had been on Evers' back.

"Our training consists of riding or power walking up and down the Canyon at a perfectly constant speed. I couldn't stop to talk with you," answered Evers, pointing to his tachometer.

"Our research indicates that you're quite a bird fancier, Mr. Evers," said the other federal agent.

"Look, I told you, I'm here training!" said Evers, getting excited. "All the bikers at Bramblewood saw me leave at sunrise. Our coach was tracking my speed through a GPS system to make sure it remained constant. All the bikers at Ashenwood saw me arrive at 2:30. Ward Thompson saw me on the trail. We passed each other at 11:00 a.m.," continued Evers.

"Is that true, Mr. Thompson? You, apparently, were hiking in the opposite direction from Ashenwood to Bramblewood," asked one of the agents.

"Not hiking—power walking!" said Ward Thompson, obviously irritated. "We cross-train for biking by power walking, to strengthen the hamstrings. I started power walking from Ashenwood at sunrise. I passed Evers at 11:00 a.m. on the trail. It was a hell of a day. I got into Bramblewood at 9:30 p.m."

"You walked all day?" asked Ravi's mother, concerned.

"Of course, we have to maintain a constant speed up and down the Canyon to build up stamina," answered Thompson, proud of his achievement.

One of the agents drew Ranger Stinson aside, coming over next to Ravi. He leaned in and whispered, "Did you see anything, Ranger Stinson? You were on duty that night."

"No, I'm sorry. I didn't see anything at all," Ranger Stinson whispered back. "Actually, I left my shift a little early. I drove out in the dark a little before sunrise, at about 5:20 a.m."

"Well, that bald eagle and its eggs are worth a lot of money, and our best guess is that one of you two gentlemen stole it," said the other federal agent, looking at Evers and then at Thompson. "Now, who wants to confess?"

Ravi listened to the questioning of Evers and Thompson while doodling with a stick in the dirt. Without looking up, he said, "Neither of them did it. The thief is…."

Who did Ravi suspect, and why?

Analysis

Often, a criminal gives himself or herself away through an inconsistency in their story. Ravi considered the following.

The most strongly established facts of the case are these:

1. The eagle could not have been stolen before sunrise. The eagle's nest is just a few minutes from the trailhead at Camp Ashenwood.

2. Ward Thompson left Camp Ashenwood (call this point A) at sunrise and power-walked at a constant speed to Camp Bramblewood (call this point B). He arrived at 9:30 p.m.

3. John Evers left Camp Bramblewood at sunrise and biked at a constant speed to Camp Ashenwood, arriving there at 2:30 p.m.

4. Thompson and Everwood crossed paths at 11:00 a.m.

Therefore, the problem facing Ravi is

> A man leaves point A at sunrise and travels to point B at a constant (unknown) speed. He arrives at point B at 9:30 p.m. A second man leaves point B at sunrise and travels to point A at a different but constant (and also unknown) speed. He arrives at point A at 2:30 p.m. The men cross paths at 11:00 a.m. What time is sunrise?

If you are able to solve this problem, you will solve the mystery as Ravi did.

Solution

There are also two parts to solving a mystery: seeing the problem that needs to be solved, and then actually solving the problem. In this case, neither is easy. It seems that there is, once again, not enough information to solve the problem.

Ravi, though, took in the facts. He realized that this problem is messy to do by algebra but could be done with a little geometry. When it appeared that he was doodling with a stick in the sand, he was actually drawing a time-distance diagram for Evers and Thompson, very much like the one shown here.

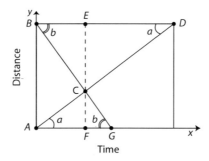

The x-axis shows the time elapsed, with the origin representing sunrise. The y-axis shows the distance traveled, from Camp Ashenwood to Camp Bramblewood. Thus, point A represents Camp Ashenwood at sunrise (the origin), and point B represents Camp Bramblewood also at sunrise. Thompson begins at A and travels along line segment AD. (The y-coordinate of point D is B, and its x-coordinate is 9:30 p.m.) Evers, meanwhile, begins at B and travels along segment BG. (Point G represents Camp Bramblewood at 2:30 p.m.) The men meet at time point C, whose time coordinate is 11:00 a.m. This is identical in the time coordinate to points E and F. The length of FG is 3.5, the difference between 2:30 p.m.—when Evers arrived at point A—and 11:00 a.m.—when the men met on the trail. Similarly, the length of ED is 10.5. To find the time of sunrise, we need to determine the length of AF (or BE).

From the diagram, we see that angle DAG equals angle BDA, because they are alternate interior angles; we'll call this angle a. Similarly, angle DBG equals angle BGA because they are also alternate interior angles; we'll call this angle b. Thus,

$$\tan a = CE/ED = CF/AF \qquad (1)$$

$$\tan b = CE/BE = CF/FG \qquad (2)$$

We can rearrange (1) to give us

$$CE/CF = ED/AF;$$

likewise, we can rearrange (2) to give us

$$CE/CF = BE/FG;$$

We can say that

$$BE/FG = CE/CF = ED/AF$$

or simply

$$BE/FG = ED/AF.$$

Now, from the diagram (and the time coordinates), we see that $BE = AF$; therefore, we have, after cross multiplying,

$$(BE)(AF) = (FG)(ED)$$
$$(AF)^2 = (FG)(ED).$$

We know the values of FG and of ED; therefore,

$$(AF)^2 = 3.5 \text{ hours} \times 10.5 \text{ hours}$$
$$AF = 6.06 \text{ hours.}$$

This is equivalent to about 6 hours and 4 minutes. Therefore, sunrise occurred 6:04 hours prior to 11:00 a.m., or at 4:56 a.m.

Obviously, Ranger Stinson was lying when he said he left the camp at 5:20 a.m., shortly before sunrise. This was about 24 minutes *after* sunrise, which gave him just enough daylight time to go to the nest, climb the rock face, and steal the eagle and its eggs.

Why he lied is not quite clear. Perhaps he was flustered and simply tripped himself up in a lie. Alternatively, he may have counted on the fact that the agents from D.C., and Ravi's family—who had slept in late—would not be oriented to the precise time of sunrise. Also, since he was out of earshot of Evers and Thompson, he thought the lie would pass unnoticed and would eliminate him as a suspect.

He was nearly correct, were it not for Ravi's doodles in the sand!

$$\frac{\binom{2n-2}{n-2}}{\binom{2n}{n}} = \frac{1}{2}\frac{n-1}{2n-1}$$

Basketball Intrigue

Ravi was dripping in sweat. He dribbled the ball up to the top of the key, nearly oblivious to the yelling of the crowd in the bleachers. Ravi was a point guard for his varsity basketball team, the Wilmont Mavericks. It was very late in the fourth quarter of the last game of the season, and the Mavericks were down 50 to 49 against the Greenville Rockets. Pat Turckle, the Mavericks' center, came out to the top of the key. Ravi eyed the clock: 20 seconds left in the game! He lobbed the ball to Pat, faked left, and exploded to the right past Pat and into the lane, with his right hand outstretched. The player guarding him got left behind as Ravi used Pat for a pick. Pat quickly passed back to Ravi. Nothing like the classic give-and-go when you need a sure basket. Ravi dribbled once and then took two steps, driving the lane for a lay-up. Ravi went up and released the ball. Out of nowhere, Ted Jones, the Rockets' star forward, jumped toward Ravi and blocked the ball with a loud slap, yelling, "REJECTED!" The crowd moaned. The Rockets recovered the ball, ran it up, and scored with two seconds left. Final score: Rockets 52, Mavericks 49.

Of course, there was no question that the Rockets were going to win the division championship. They had by far the best record in the league. The Mavericks were second, and

although they had no shot at the Division Title, Ravi hoped
that they could beat the Rockets in the last game of the
season. However, everyone in the division knew that when
Jones and Sadowski, the Rockets' star center, were playing
together on the same team, they were essentially unbeat-
able. Ravi played in a unique league where each meeting
of opposing schools actually consisted of four games in a
round-robin tournament. Ten players were on each school
squad. Prior to each tournament, the coaches divided the
players at random, by picking names out of a hat, into two
teams of five, team A and team B. Those two teams would
then play a round-robin tournament with the two teams
similarly chosen from the opponent school, for a total of four
games. If one school won 3 or 4 of the games, that school
was declared the winner of the tournament. If the schools
split the games 2–2, the tournament was declared a tie.

This year, the Rockets had a roster full of excellent
players. However, they had two especially great players,
Ted Jones and Michael Sadowski. When they got picked
onto the same team, they could not be defeated. That combination had not lost once all season! This year, the Rockets
had been lucky. Jones and Sadowski seemed to get paired
together onto the same five-man team a lot. When parents
from the other teams grumbled to the Rockets' Coach, John
Durski, he would just shrug his shoulders and say, "They're
just as likely to be together as apart. Anyway, better to be
lucky than good. This year, we're both!"

Soon, the crowd dissipated, and the gym was nearly
empty. The Rockets team headed out for dinner. Ravi decided
to hang around school until the Athletic Awards ceremony
at 7:00 p.m. It was only two hours, and Ravi didn't feel
like biking home. Besides, he wanted to do some practice

problems for the AMC 12 math test, and a little quiet time would be a good chance to do just that. Ravi walked toward the doors of the gym, passing by the Officials' Table, where the trophy for Division Champion stood, already inscribed with the name of the Greenville Rockets, ready to be presented later that evening. As he walked by the table, he stopped to look over the game stats from this season. The Rockets certainly had an impressive record. They had won 17 out of 20 round robin tournaments against the other teams in the division. As Ravi perused the team rosters, he casually noted that in 14 of the 20 tournaments, Jones and Sadowski had been paired onto the same team. Ravi walked out of the gym and toward the library. "Time to hit some problems," he thought, musing over the season he had enjoyed so much.

At 7:00 p.m., Ravi walked back to the gym. The bleachers were filled with students and parents both from Greenville and from Wilmont. The Rockets were in uniform, and Coach Durski had changed into a suit. Mr. Arvin Wilson, head coach of the Southwest Division, walked over to the podium that had been set up in the middle of the gym and began talking: "Attention please! Good evening parents and students. Well, another season has ended, and it is time to crown our Division Champs. I ask Coach Durski to please come to the podium, and I ask the Greenville Rockets to please stand up and be recognized!" The Rockets rose to their feet to the applause of the parents and students. Coach Durski walked over to the podium and raised his hand to acknowledge the crowd, getting ready to speak. Meanwhile, Head Coach Wilson walked over to the Officials' Table to pick up the trophy for its official presentation. Ravi strode over to the table to meet him.

"Excuse me, Coach Wilson," said Ravi.

"Yes, may I help you, young man?" answered Coach Wilson.

"Yes. Do not present the trophy tonight. I think there's been some cheating."

Of course, Coach Wilson did not listen to Ravi, and the presentation went on as scheduled. However, after the ceremony was over, Ravi showed Coach Wilson the calculations he had been working on in the library after the game. Based on that, an official investigation was launched, and the assistant coach of the Rockets admitted to conspiring with Coach Durski to rig the team roster so that Jones and Sadowski could get on the same team more than simple chance would have allowed. The Rockets were stripped of the trophy, and the Mavericks, as the second place team, were crowned the new champion. There was no pomp or ceremony, but Ravi was hailed by the team as "most valuable player." This was not so much for his basketball skills as for his mathematical prowess. Why did Ravi suspect cheating?

Analysis

The problem in this case can be stated in two parts:

1. If we have ten players on the Rockets, who are divided into two five-person teams, A and B, what is the likelihood that two particular players (e.g., Jones and Sadowski) will end up on the same team?
2. Given the answer to Part 1, what is the likelihood that this would happen in as many as 14 out of 20 tournaments?

Solution

Part I

To make the problem slightly easier to discuss, let us number the players 1 through 10 and label Jones and Sadowski as player 9 and player 10, respectively. Now, if we divide these ten players randomly into two five-player teams A and B, what is the likelihood that players 9 and 10 are on the same team?

First, let us begin by finding the total number of ways that we can divide the players into two teams A and B. This is simply the number of ways we can pick 5 players out of 10 for team A. The remaining 5 players then become team B.

Let's start with a smaller example. Suppose that an ice-cream store has eight flavors of ice cream and sells a three-scoop banana split. If each scoop is a different flavor, how many different banana splits can you order?

Let's start by picking one flavor at a time. There are eight choices for the first flavor; then there are only seven flavors left for the second choice, and finally only six for the third. So, there are

$$8 \times 7 \times 6 = 336$$

ways to pick three flavors—in order! To the clerk, it doesn't matter if you order a vanilla-chocolate-strawberry split or a strawberry-vanilla-chocolate split. (They still scoop the same ice cream.) The number above contains duplicate orders; we want to remove the repeats. Let's consider three flavors—C, S, and V—and the number of different ways to list them in order:

C S V
C V S
S C V
S V C
V C S
V S C

Or, along the lines of our previous argument, $3 \times 2 \times 1 = 6$. There are a total of six ways to ask for the same banana split. So, in the 336 ways to say the ice cream names in order, we are repeating one order six times. Therefore, there are

$$\frac{8 \times 7 \times 6}{3 \times 2 \times 1} = \frac{336}{6} = 56$$

different banana splits you could order.

The process that we just went through can be summarized by an equation that uses factorials. As you may know, the factorial $n! = 1 \times 2 \times ... \times (n - 1) \times n$. In combinatorics (the field of mathematics that deals with counting and combinations), for any n objects, the number of ways to choose k out of those n objects is called $C(n,k)$, or "n choose k." This is usually written as follows:

$$\binom{n}{k}, \text{ which stands for } \frac{n!}{(n-k)!k!}.$$

Therefore, if an ice-cream store has eight flavors of ice cream, the number of ways to pick three flavors for a banana split is $C(8,3)$ or $\binom{8}{3}$, which is calculated as follows:

$$\frac{8!}{(8-3)!3!} = \frac{8!}{5!3!} = \frac{8 \cdot 7 \cdot 6 \cdot \cancel{5} \cdot \cancel{4} \cdot \cancel{3} \cdot \cancel{2} \cdot \cancel{1}}{(\cancel{5} \cdot \cancel{4} \cdot \cancel{3} \cdot \cancel{2} \cdot \cancel{1})(3 \cdot 2 \cdot 1)} = \frac{8 \cdot 7 \cdot 6}{3 \cdot 2 \cdot 1} = 56.$$

You can see that this matches our calculations from finding the value "the long way."

Using this equation, let's go back to the basketball problem. The total number of ways to make two teams A and B is

$$\binom{10}{5} = \frac{10!}{(10-5)!5!} = \frac{10!}{5!5!} = 252\,.$$

This counting distinguishes the order of the teams, so that if players 1–5 are team A and players 6–10 are team B, this is counted as different from players 6–10 being team A and players 1–5 being team B.

Now, to figure out the probability that players 9 and 10 are on the same team, let us start by calculating the probability that they are together on team A.

If we assign players 9 and 10 to team A, there are three more spots to fill and eight players left to fill them from. Therefore, there are

$$\binom{8}{3} = 56$$

ways to make team A with players 9 and 10 on it. Similarly, there are 56 ways to make team B with players 9 and 10 on it. Therefore, the probability of players 9 and 10 being together on either team A or team B is

$$\frac{56+56}{252} = \frac{4}{9}\,.$$

Notice that this is a bit different from the claim of Coach Durski (and possibly our intuition) that his two best players are as likely to be together as apart. Actually, the probability of them being together on a team is 4/9 (we'll call this probability p), while the probability that they are on different teams is slightly higher at 5/9 (we'll call this q, which equals $1-p$).

Part 2

Now that we have calculated that p = 4/9, we need to ask: if the players are divided into teams A and B for 20 tournaments, how likely is it that players 9 and 10 end up playing together on the same team 14 out of the 20 times?

This is a problem that can be solved by using the binomial distribution. Let us say that an event has only two possible outcomes, like the flip of a coin, either H or T. If outcome H occurs with probability p, then T must occur with probability $q = 1 - p$. Now, let's say that we repeat the event five times. One possible outcome set is H H H T T. The probability of this outcome set occurring is

$$p \cdot p \cdot p \cdot (1-p) \cdot (1-p) = p^3 (1-p)^2.$$

Now, let us look at another possible outcome set for the five trials: H T T H H. The probability of this outcome set occurring is

$$p \cdot (1-p) \cdot (1-p) \cdot p \cdot p = p^3 (1-p)^2.$$

In fact, any outcome set that has 3 H's and 2 T's in some order will have this same probability. Now, if we ask how *likely* it is that for five events we end up with 3 H's and 2 T's, regardless of order, we need to multiply the probability $p^3(1 - p)^2$ by the number of ways we can order 3 H's and 2 T's in a set of five outcomes. This is just the number of ways we can choose three outcomes out of the five to be H (the remaining two will be T by default). Thus, this is $C(5,3)$, and the probability of an outcome set of 3 H's and 2T's is

$$\binom{5}{3} p^3 (1-p)^2.$$

Now, let us say that an event has only two possible outcomes, H and T, with probabilities p and $(1 - p)$, respectively. Then, if we repeat the event n times, then the probability $P(k)$ that k of these turn out to be H is

$$P(k) = \binom{n}{k} p^k (1-p)^{n-k}.$$

For our problem, the outcome we are interested in is that Jones and Sadowski get picked to be on the same team. (This is analogous to H above, with T analogous to the two players getting picked for different teams.) The probability of this outcome is $p = 4/9$. Therefore, the likelihood that for $n = 20$ games they would be together $k = 14$ times is

$$P(14) = \binom{20}{14}\left(\frac{4}{9}\right)^{14}\left(\frac{5}{9}\right)^{20-14}.$$

To calculate the actual value, we would rewrite the expression as follows:

$$P(14) = \left(\frac{20!}{14!\,6!}\right)\left(\frac{4}{9}\right)^{14}\left(\frac{5}{9}\right)^{6}.$$

This can be figured out on a calculator or PC and has a value of 0.0134, or 1.3%.

More generally, the probability that they would be together k out of 20 times is

$$P(k) = \binom{20}{k}\left(\frac{4}{9}\right)^{k}\left(\frac{5}{9}\right)^{20-k}.$$

These values are seen in the graph on the next page.

We notice that it is unlikely that they would be paired together very few times or very many times. The highest probability is that they would have been paired together 9 out of 20 times, with a probability of 0.1768, or close to 18%; being paired together 8 or 10 times are only slightly less

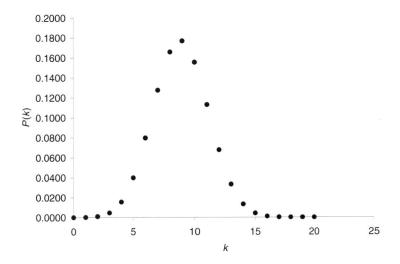

likely, and the chances drop to both sides of these high values. If we sum all of these probabilities for the possible values of k (i.e., 0–20), they total to 1 (or 100%), as we expect:

$$\sum_{k=0}^{20} \binom{20}{k} \left(\frac{4}{9}\right)^k \left(\frac{5}{9}\right)^{20-k} = 1.$$

Now, to determine whether it was simply chance that Jones and Sadowski got paired together 14 times out of 20, it is not fair for us to just look at $P(14)$. What we are really asking is how likely it is that they would be paired together *at least* 14 times, so we need to sum the probabilities for them getting paired together 14, 15,..., and 20 times,

$$\sum_{k=14}^{20} \binom{20}{k} \left(\frac{4}{9}\right)^k \left(\frac{5}{9}\right)^{20-k} = 0.0189512.$$

Thus, the likelihood that Jones and Sadowski would be paired together as many as 14 times out of 20 is less than 2%.

On this basis, Ravi concluded that some cheating had probably occurred in the pairing of the players, to assure that Jones and Sadowski would end up together. Furthermore, he could make that statement with greater than 98% likelihood of it being true, and that was sufficient enough to launch an investigation.

Extension

At least two other interesting problems present themselves, and it would be nice to consider them.

1. Let us assume that there are three superstar players out of the ten. If the ten players are randomly divided into two teams of five, what is the likelihood that all three players would end up on the same team?

As we have seen, there are 252 ways, $C(10,5)$, to divide the players into two teams, where the order of the teams matters. Now, let us assign the three top players to team A. There are two spots left and seven players left to fill them. This is "7 choose 2": i.e.,

$C(7,2) = \dbinom{7}{2} = 21$ ways to make team A with the three top players on it.

Similarly, there are 21 ways to make team B with the three top players on it. Therefore, the likelihood that the three superstars end up together on either team is $(21 + 21)/252 = 1/6$, or about 16.7%.

2. Now, let us assume that the Rockets have 15 players, which will be randomly divided into three teams of five. What is the probability that Jones and Sadowski end up on the same team now?

First, let us determine how many ways there are to divide the 15 players into 3 teams of 5. There are $C(15,5)$ ways to pick five players for team A. Now, for each of those teams, there are ten players left from which to pick the five for team B, i.e., $C(10,5)$ ways to make team B for each team A. This leaves five players who by default make up team C. Thus, the total number of ways is

$$C(15,5) \cdot C(10,5) = \binom{15}{5} \cdot \binom{10}{5}$$

ways. If we expand this into factorial notation, we get

$$\frac{15!}{10!5!} \cdot \frac{10!}{5!5!} = \frac{15!}{(5!)^3}$$

different ways to divide the 15 players into three teams, where the order of the teams matter.

As an aside, if the order of the teams does not matter, but simply which players are grouped with which others, we need to divide the above number by the number of ways we can order teams A, B, and C—i.e., 3!—to get

$$\frac{15!}{3!(5!)^3} .$$

This works out to be 126,126 different ways.

In the same fashion, we can generalize that the number of ways of dividing nk objects into k groups of n objects each is

$$\frac{(nk)!}{k!(n!)^k} .$$

To calculate the number of ways Jones and Sadowski can be together on team A, we first assign them to team A. Then, that leaves three spots to fill and 13 players to pick from, i.e., $\binom{13}{3}$. Now, for each of these choices for team A, there are ten players left, and hence there are $\binom{10}{5}$ ways of making team B, with team C being thus automatically determined. Therefore, there are

$$\binom{13}{3}\binom{10}{5} = \frac{13!}{3!10!} \cdot \frac{10!}{5!5!} = \frac{13!}{3!(5!)^2}$$

ways of making the three teams with Jones and Sadowski together on team A. Analogously, there are an identical number of ways of having them together on team B or on team C; i.e., $\frac{3(13!)}{3!(5!)^2}$ ways of having Jones and Sadowski together on any of the teams. Thus, the probability that they would end up together on one of the teams is

$$\frac{3(13!)}{3!(5!)^2} \div \frac{15!}{(5!)^3} = \frac{3(13!)}{3!(5!)^2} \cdot \frac{(5!)^3}{15!} = \frac{3(13!)(5!)}{3!15!},$$

which equals 2/7, i.e., approximately 0.2857, or about 29%.

Lastly, it is nice to note that if we assign Jones and Sadowski to team B first, we get the same result. There are then 13 players from whom to choose 5 to put on team A, or $\binom{13}{5}$ ways. Then, there are eight players left. There are now three slots to fill on team B, and so there are $\binom{8}{3}$ ways to make team B for each team A. Team C is made up of the five remaining players and hence is automatically determined. Thus, if we assign Jones and Sadowski to team B first, there are

$$\binom{13}{5}\binom{8}{3} = \frac{13!}{5!8!} \cdot \frac{8!}{3!5!} = \frac{13!}{3!(5!)^2}$$

ways to make the teams, just as claimed.

If we assign Jones and Sadowski to team C, then once again there are $\binom{13}{5}$ ways to make team A. Once five players are chosen, this leaves $\binom{8}{5}$ ways to make team B. Once those five players are chosen, the remaining three players fill out team C. Thus, there are indeed

$$\binom{13}{5}\binom{8}{5} = \frac{13!}{5!8!} \cdot \frac{8!}{3!5!} = \frac{13!}{3!(5!)^2}$$

to make the three teams with Jones and Sadowski on team C.

The Moon Rock

Ravi watched the basketball sail over the backboard, "Nice try, Dad. Maybe a little softer will do it."

Ravi's father was not a good basketball player. He shot in the awkward two hands from the chest style used by adults who have never played basketball before. On top of that, he was not a naturally coordinated person. Nonetheless, being a busy district attorney, he always made a conscious effort to do father-son things with Ravi on the weekends, like "shoot hoops." Ravi would have been perfectly content to just talk with his father or play chess, but he went along to make his father happy.

As the father dribbled the ball, he said, "The police have a man in custody who apparently knows you. I talked to him briefly yesterday. We are going to indict him for attempted robbery."

"And he knows me?" asked Ravi incredulously.

"Yes," replied Ravi's father, "he is not your typical criminal and this is not your typical crime. Apparently, he was going to steal a moon rock from the Chicago Museum of Science. He is a night watchman there. His name is..."

"Not George Davis!" exclaimed Ravi.

"So you do know him," remarked the father.

"Yes, Dad, I know him well. I can't believe Mr. Davis would steal from the museum."

"Well, the police got an unexpected tip from Mr. Davis' partner who had changed his mind the night before the heist. It seems that the moon rock has a fairly unsophisticated alarm system which consists of a weight sensor in the platform on which it sits. The sensor itself is fairly sensitive, to within 1 milligram or so, but all one has to do is to know the weight of the moon rock beforehand and simply put precisely the same weight on the platform as they remove the moon rock."

"So what does that have to do with Mr. Davis?" asked Ravi.

"His partner told us that they were planning to steal the moon rock at night. It is worth just under $1,000,000 on the black market. The caller told us that we would find metal weights in Mr. Davis' bag, which he planned to use to keep the weight sensitive alarm from going off," said Ravi's father. "So the police went to the museum and found a bunch of metal washers in Mr. Davis' bag, just like the caller said. They arrested him and brought him downtown."

"Well, I just don't believe it," exclaimed Ravi.

"Why not?" asked his father. "How do you know Mr. Davis anyway?"

"Let me ask you, Dad; what do you know of Mr. Davis?" replied Ravi.

"I know he has been the night watchman at the museum for thirty-four years. I know that he is an old man who has spent his whole life working for minimum wage who saw an easy get-rich-quick opportunity," said Ravi's father, in the manner of a man who had dealt with many such petty criminals in the past.

"Well, Dad, that doesn't sound at all like the Mr. Davis I know. As you know, I'm sort of a regular at the science museum, and over the last few years during my visits there, I think I've gotten to know Mr. Davis pretty well. I can honestly say he is someone that I respect and trust," said Ravi.

"Respect?" inquired Ravi's dad, astonished, since that was not a way he heard Ravi describe many people.

"Yes, Dad. Mr. Davis was a graduate student in astronomy at Berkeley in the late sixties. He was in love with a girl who was working on her PhD in mathematics. Shortly before they got married, she was diagnosed with leukemia. He insisted on marrying her, and she died just a few months later. He told me how much they used to love walking on campus or going to the ocean and spending evenings looking at the stars and talking about astronomy, mathematics, and philosophy. Mr. Davis has long believed that the universe is completely purposeful, without mistakes or coincidences—that it is as perfectly balanced as the mathematical equations that his bride so loved. After she died, he realized that he had no interest in finishing his PhD and that all he really wanted to do was to learn and to read. Therefore, he left Berkeley, came to Chicago, and took the job of night watchman at the science museum so that he could still be close to science and have all the time that he wanted to read."

"I didn't realize that about Mr. Davis, Ravi. However, it does not change the evidence against him. He didn't bother to deny the charges. When I was introduced to him as the district attorney, he apparently recognized my name and perhaps a resemblance between us and simply asked me if I knew a young man named Ravi. When I replied that you were my son, he said that he thought you were a very fine young man and he asked me to say hello to you."

"Dad, we have to go down to the police station and get to the bottom of this," said Ravi determinedly.

By now, his father had learned to trust both Ravi's instinct and his intellect. If Ravi had serious doubts about this man's guilt, then the matter needed further attention.

Ravi and his father arrived at the Chicago metropolitan police station. They walked to the office of Detective Myers. Ravi's father greeted the detective and said, "Detective Myers, this is my son, Ravi. With your permission, I would like the three of us to go talk to Mr. Davis again."

Detective Myers eyed Ravi with some suspicion, cocking an eyebrow. He obviously did not appreciate a teenager wasting his time. However, he had enough respect for Ravi's father to confine his objections to that single gesture.

"Fine, let's go," he said curtly.

They followed him to the holding cell where Mr. Davis was being kept until Monday's indictment.

"Well, hello, young man," said Mr. Davis, addressing Ravi as the trio entered his cell.

"Hello Mr. Davis. What's this all about?" asked Ravi, concerned.

"Apparently, after thirty-four years, the Chicago Museum of Science has decided that I'm a thief," said Mr. Davis wistfully.

"I know you're not a thief," said Ravi. "Please tell me what is going on."

Mr. Davis looked up and took a deep breath, "Actually, it's very simple. I think I'm being framed. Mr. Levy, the museum director, has been after me to retire for a year. He wants to give the job to his son-in-law."

"What about the weights they found in your bag, Mr. Davis?" asked Ravi.

"I don't know anything about any weights," replied Mr. Davis. "The first time I saw them was when the police showed up and searched my bag. However, I presume that you think that the weights are there to deactivate the alarm in the platform which holds the moon rock. That would be difficult, because the alarm has a tolerance of 1 mg under or over the weight of the moon rock. Unless the weight is replaced precisely as the moon rock is lifted, the alarm would sound."

"Interesting, Mr. Davis. How much did the metal washers you found in Mr. Davis' bag weigh, Detective Myers?" asked Ravi, turning to the detective.

The detective pursed his lips, now looking truly annoyed. He pulled a small notebook out of his pocket.

"238 grams. We weighed them on the scale used for drugs in the evidence room," replied the detective.

"And how much does the moon rock weigh?" continued Ravi.

The detective flipped through his notebook and looked up, saying, "I don't know. We didn't pick it up to weigh it."

"It weighs precisely 95 grams. It was picked up by an astronaut from a small crater in the Sea of Tranquility. It is 38% olivine, 27% pyroxene, 19% plagioclase, and 16% ilmenite by weight," said Mr. Davis, to the astonishment of everyone present.

"Detective Myers, my father mentioned that the weights found in Mr. Davis' bag were in the form of metal washers. Were they all the same size? How much did each washer weigh?" asked Ravi.

"Once again, I don't have that written down," said the detective, now visibly showing his irritation, "but what difference does that make? He had more than enough weight to make 95 grams."

"As the district attorney on this case, Detective Myers, I would like to examine the washers. Could you please bring them and your scale so that my son can check them," said Ravi's father sternly, bothered by the disrespect the detective was showing to Ravi.

The detective complied and brought a plastic bag containing the metal washers, and the sensitive scale used to weigh drugs from the evidence room. Ravi spread the washers out on the table. There were only two sizes of washers. There were 12 small ones and 10 slightly larger ones. Ravi picked up one of the small washers and put it on the scale.

"Nine grams," Ravi announced.

He repeated the procedure with the large washer. "Thirteen grams," he stated.

"Hmmm," Ravi said, closing his eyes as he did when he thought deeply.

After a few moments of concentration, he opened his eyes and looked at Mr. Davis, who was smiling.

"That was a close one, Mr. Davis," said Ravi, addressing his old friend.

"A little too close, young man. Quite a coincidence, don't you think, Ravi?" said Mr. Davis, still smiling at Ravi.

"Now, Mr. Davis, you know that there are no coincidences in the universe," grinned Ravi, winking at Mr. Davis.

Ravi turned to his father and Detective Myers, saying, "Obviously, Mr. Davis is innocent."

♀

Both Ravi's father and Detective Myers could see that Mr. Davis obviously knew the weight of the moon rock. Ravi figured out that

95 grams could not be made from a combination of 9-gram and 13-gram washers. In other words, there are no nonnegative integers x and y such that $9x + 13y = 95$. Therefore, Mr. Davis was released.

That is the easy part of the story. The interesting part revolves around the conversation between Ravi and Mr. Davis at the end. What do you think that meant?

Analysis

This story leads us directly to a very interesting problem. If we have two positive integers a and b, what numbers can be made as linear combinations of these integers, i.e., $c = ax + by$ where x and y are nonnegative integers?

To make this concrete, the problem can be restated as what weights can be made from a combination of washers that weigh a grams and b grams, or what postage denominations can be made from stamps that are worth a cents and b cents, etc.?

In our case, $a = 9$ and $b = 13$, and we see that the number 95 could not be made as a linear combination of these. The conversation between Ravi and Mr. Davis seemed to imply that there was something special about this number 95—but what?

Solution

Where do we begin investigating such a problem? Usually, if there is no obvious line of attack, it is best just to experiment a little bit to get familiar with the problem, preferably with some easy examples. For the sake of simplicity, let's say that a is always less than b.

Let us pick $a = 4$ and $b = 5$ and make a chart of which numbers can be made as a linear combination of these (we'll call these *good* numbers) and which numbers cannot be made in such a way (we'll call these *bad* numbers).

$a = 5, b = 4$

Number	0	1	2	3	4	5	6	7	8	9	10	11	12	13	14	15	16	17	18
Good	*				*	*			*	*	*		*	*	*	*	*	*	*
Bad		x	x	x			x	x				x							

Let us look at the first few good numbers:

$$0 = 0 \times 4 + 0 \times 5,$$
$$4 = 1 \times 4 + 0 \times 5,$$
$$5 = 0 \times 4 + 1 \times 5,$$
$$8 = 2 \times 4 + 0 \times 5,$$
$$9 = 1 \times 4 + 1 \times 5,$$
$$10 = 0 \times 4 + 2 \times 5,$$
$$12 = 3 \times 4 + 0 \times 5,$$
$$13 = 2 \times 4 + 1 \times 5,$$
$$14 = 1 \times 4 + 2 \times 5,$$
$$15 = 0 \times 4 + 3 \times 5,$$
$$16 = 4 \times 4 + 0 \times 5,$$
$$17 = 3 \times 4 + 1 \times 5,$$
$$18 = 2 \times 4 + 2 \times 5.$$

Now, with just this easy exercise, a couple of interesting points arise that may be useful later. First, we can speculate that 11 will be the last bad number, and that all numbers after it will be good numbers. Thus, let us call 12 the first *permanently good* number. How do we know that 11 will be the last bad number? Well, once we have found four good numbers in a row, we know that all of the rest of the numbers will be good numbers! This is because $a = 4$, and so we can generate a new set of four good numbers simply by adding one more a to each of our prior four good numbers, to infinity. Thus, $13 = 2 \times 4 + 1 \times 5$, while $17 = 3 \times 4 + 1 \times 5$; $14 = 1 \times 4 + 2 \times 5$, while $18 = 2 \times 4 + 2 \times 5$, and so on. Therefore, our example suggests the following:

1. For any a and b, there may always be a last bad number.

2. We will find this number once we find a string of a ($a < b$) good numbers in a row. The beginning of that string is the first permanently good number (such as 12 in this case).

These are not conclusions, of course, but just conjectures to focus on. If we start thinking about the first point, we quickly realize that it is not true for some choices of a and b. Suppose that a and b have a common factor f ($f \neq 1$). Then, all linear combinations of a and b will have to be multiples of f. Thus, if $a = mf$ and $b = nf$, then $c = ax + by$ leads to $c = mfx + nfy$ and then $c = (mx + ny)f$; i.e., c is always a multiple of f. Therefore, any numbers that are not multiples of f will be bad numbers (giving us an infinite number of bad numbers). This means that there will never be a permanently good number.

Thus, we will modify our first point to the following:

1. For any relatively prime integers a and b, there may always be a last bad number.

Now, we ask the reader to repeat the process above for a few more cases to see if he or she can spot a pattern. For a and b relatively prime, can we always find a string of a consecutive good numbers, thus finding the last bad number and the first permanently good number? Try it for the following pairs of numbers:

$$a = 2, b = 5;$$
$$a = 3, b = 5;$$
$$a = 5, b = 6;$$
$$a = 5, b = 7;$$
$$a = 3, b = 4;$$
$$a = 4, b = 7;$$
$$a = 4, b = 9.$$

That will take a bit of work, but sometimes it's hard to solve a math problem without getting your hands dirty! Compare your results to the table below.

a	2	3	4	5	5	3	4	4
b	5	5	5	6	7	4	7	9
First permanently good number	4	8	12	20	24	6	18	24

Looking at this table, is it possible to spot a pattern? If we are lucky, we may eventually hit upon a pattern that seems to agree with all the entries in our table thus far. Can you see it?

First permanently good number = $(a - 1)(b - 1)$.

Now, we actually have a concrete conjecture, albeit one obtained by trial and error:

The last bad number is $(a - 1)(b - 1) - 1 = ab - (a + b)$.

Now, we are ready to search for a proof. We will show two parts:

(i) For a and b relatively prime, $ab - (a + b)$ is always bad.

(ii) Assuming that $a < b$, the next a consecutive integers after $ab - (a + b)$ will all be good numbers. These are

$$ab - a - b + 1 = (a - 1)(b - 1),$$
$$(a - 1)(b - 1) + 1,$$
$$(a - 1)(b - 1) + 2,$$
$$\cdots$$
$$(a - 1)(b - 1) + (a - 1) = (a - 1)b.$$

Proof of (i). We will do this proof by contradiction, assuming that $ab - (a + b)$ is a good number and showing that this leads to a contradiction. We already know that a and b are relatively prime positive integers.

Thus, by assumption,

$$ab - (a + b) = ax + by, \text{ where } x \text{ and } y$$
$$\text{are both nonnegative integers.}$$

Considering this assumption, we claim that $y < a - 1$. To prove this claim, we note that $ab - (a + b) < ab - b$, or $ab - (a + b) < (a - 1)b$; therefore, y has to be less than $a - 1$. Otherwise, if $y \geq a - 1$, then $ax + by$ would be greater than $ab - (a + b)$.

Now we rearrange our equation $ab - (a + b) = ax + by$ to get

$$x = \frac{ab - a - b - by}{a};$$

This can be further rearranged to give

$$x = \frac{b(a - 1 - y) - a}{a}.$$

At this point, we observe that if $(w - a)/a$ is an integer, then a must divide w. Thus, for x to be an integer, a must divide $b(a - 1 - y)$. Since a and b are relatively prime, a cannot divide b, which means that a must divide $a - 1 - y$. Since $y < a - 1$, the term $a - 1 - y$ is greater than zero. However, it is also less than a, meaning that a cannot divide $a - 1 - y$. Therefore, there can be no such x to complete the assumed linear combination above. Thus, by contradiction, $ab - (a + b)$ is always bad.

Proof of (ii). Immediately, we see that the last number in the list, $(a - 1)b$, is always a good number, because it is achievable by setting $x = 0$ and $y = (a - 1)$ in $ax + by$. Therefore, we really only need to check the $a - 1$ numbers $(a - 1)(b - 1), (a - 1)(b - 1) + 1, \ldots , (a - 1)(b - 1) + (a - 2) = (a - 1)(b - 1) + (a - 1) - 1 = (a - 1)b - 1$.

This same list of $a - 1$ numbers can be rewritten as $(a - 1)b - k$, where k is, in turn, each of the numbers 1, 2, ..., $a - 2, a - 1$.

Once again, assuming that a and b are relatively prime and that $a < b$, we will write b as a multiple of a plus some remainder r:

$$b = qa + r, \text{ where } 1 \leq r \leq a - 1 \text{ and } q \text{ is an integer.}$$

We note that a and r are also relatively prime; otherwise, if $a = md$ and $r = me$, then $b = qmd + me = m(qd + e)$. Thus, b and a would both have a common factor m and would not be relatively prime.

To begin our proof, we will start by looking at another list of $a - 1$ numbers: $r, 2r, \ldots (a - 1)r$. Clearly, none of these numbers is divisible by a (since a and r are relatively prime and each term is of the form ir where $i < a$). The fact that

a and r are relatively prime also means that none of these numbers have the same remainder when divided by a. If they did, then for some $i < j$ the number $(j - i)r$ would be divisible by a. Since $(j - i)r$ is in our list of numbers, we have established that this is impossible. (If $ir = sa + z$ and $jr = ta + z$, then $(j - i)r = (t - s)a$.)

Now, we know that if we divide an integer by a, there are a possible remainders: $0, 1, 2, ..., a - 1$. If that integer is relatively prime to a, then zero cannot be a remainder and there are only $a - 1$ possibilities. Since the $a - 1$ numbers in the list $r, 2r, ...(a - 1)r$ each have a unique remainder when divided by a, each number from 1 to $a - 1$ will occur exactly once in the list of those remainders.

Therefore, for any k, $1 \leq k \leq a - 1$, there is some value mr from the above list that leaves remainder k, i.e., $mr = na + k$.

Now, we write $(a - 1)b - k$ as $(a - 1)(qa + r) - k$ and proceed to show that this is always a good number. First,

$$(a - 1)(qa + r) - k = (a - 1)qa + (a - 1)r - k.$$

Now, we add and subtract mr from the right side to get

$$(a - 1)(qa + r) - k = (a - 1)qa + (a - 1 - m)r + mr - k.$$

Again to the right side, we add and subtract mqa to get

$$(a - 1)(qa + r) - k = (a - 1 - m)qa + (a - 1 - m)r + mr - k + mqa.$$

Rearranging the right side, we get

$$(a - 1)(qa + r) - k = (a - 1 - m)(qa + r) + mr - k + mqa.$$

Now, we substitute $mr = na + k$ and get

$$(a - 1)(qa + r) - k = (a - 1 - m)(qa + r) + (n + mq)a.$$

All that is left is to remember that $b = qa + r$, and we can substitute this back to get

$$(a - 1)b - k = (a - 1 - m)b + (n + mq)a.$$

This is valid for each value of k, and it shows that each of our $a - 1$ numbers $(a - 1)b - k$, $1 \leq k \leq a - 1$, is a good number. That is true because each number can be rewritten as a linear combination of a and b, $ax + by$, where $y = (a - 1 - m)$ and $x = (n + mq)$.

Adding to this that $(a - 1)b$ is a good number, we have proven that there are a consecutive good numbers, $(a - 1)(b - 1)$, $(a - 1)(b - 1) + 1$, ... , $(a - 1)b$.

Thus, we have also proven that $(a - 1)(b - 1) - 1 = ab - (a + b)$ is the last bad number and that $(a - 1)(b - 1)$ is the first permanently good number, as long as a and b are relatively prime!

Now, returning to our story, we can finally appreciate the last exchange between Ravi and Mr. Davis. The moon rock weighed exactly 95 grams, and the washers planted on Mr. Davis weighed 9 and 13 grams, respectively. We see that for $a = 9$ and $b = 13$, the last bad number is $(a - 1)(b - 1) - 1 = 8 \times 12 - 1 = 95$. This is the last weight that could not have been made as a combination of 9-gram and 13-gram washers. Had the moon rock weighed anything more, regardless of what value the weight had above 95 grams, that number could have been made as a combination of the washers, and there would have been no way to prove Mr. Davis' innocence!

Extension

Having solved this problem, it is natural to wonder what weights can be made from three different washers, weighing a, b, and c grams, respectively (again when combinations of the available washers are put on a single pan scale).

More mathematically, if we have positive integers a, b, and c, what numbers can and cannot be made as linear combinations of a, b, and c; i.e., $d = ax + by + cz$?

This simple extension actually makes the problem very hard, and there is no elementary solution. However, for fun, we can illustrate a trick that can be used in some special cases.

Once again, we note that if all three numbers have a common factor f, then any linear combination will also have to be a multiple of f; therefore, all numbers that are not multiples of f will be bad, and there will never be a last bad number. So, for example, a, b, and c cannot be all even, etc.

Let's look at an example that we can solve, with some ingenuity. If $a = 12$, $b = 20$, and $c = 33$, what is the last bad number? In other words, what is the largest integer that cannot be represented as $12x + 20y + 33z$?

Let us rewrite the equation as $3(4x + 11z) + 20y$. Now, let's look at the term in brackets, $4x + 11z$. This is a linear combination of $a = 4$ and $b = 11$; from our prior discussion, we know that the first permanently good number is $(a - 1)(b - 1) = 10 \times 3 = 30$. Therefore, all numbers from 30 upward can be represented by the term in brackets. This will now be replaced with $(t + 30)$, and the equation can be rewritten as

$$3(t + 30) + 20y = 3t + 20y + 90.$$

Now, we look at $3t + 20y$; again, we see this as a linear combination of 3 and 20, and we know that all numbers starting from $(3 - 1)(20 - 1) = 38$ can be made from it. Therefore, $3t + 20y$ can be replaced by $s + 38$, and our equation becomes

$$s + 38 + 90 = s + 128.$$

Therefore, all numbers starting from 128 can be represented as $12x + 20y + 33z$, and 127 is the last bad number for $a = 12$, $b = 20$, and $c = 33$.

This little trick is basically applying the result derived in the solution twice in a row. But wait a minute; let's look at our equation again to make sure that there is no trickery here:

$$12x + 20y + 33z$$

Let's say, instead, that we had decided to group $12x + 20y$ together first and to write the equation as $4(3x + 5y) + 33z$. What happens now? Let's find out.

Looking at the term in brackets, $3x + 5y$, we know that all numbers beginning from $2 \times 4 = 8$ can be represented, and we rewrite the equation as

$$4(t + 8) + 33z = 4t + 33z + 32.$$

Now, looking at $4t + 33z$, we know that all integers from $3 \times 32 = 96$ can be represented. Therefore, all integers from $96 + 32 = 128$ can be represented, and once again 127 is the last bad number.

Of course, had the result been different, our method would have proved invalid. Keep in mind, though, that we got the answer by finding an algorithm to generate the solution. That is quite different than having an easy formula as we did in the previous section, but not bad for a day's work!

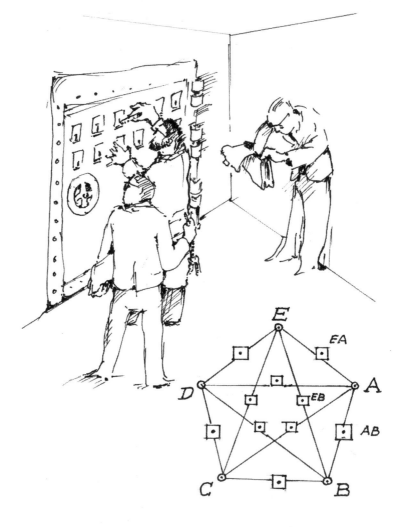

A Theft at Dubov Industries

Ravi's school day began like any other, in logic class. Today, he had a test, and it consisted of only two multiple choice questions worth fifty points each. Unfortunately for Ravi, the copying machine had malfunctioned, so that the second question on his paper was entirely unreadable. All he could read were the answer choices beneath the question:

A) All of the below.

B) None of the below.

C) All of the above.

D) One of the above.

E) None of the above.

F) None of the above.

Little did Ravi know that this would only be his first challenge of the day. When he returned home, he found a large black limousine parked in the driveway of his house.

"You are Ravi, yes?" asked the tall, neatly-dressed man who stepped out from the back of the limousine, speaking in a thick Russian accent.

"Yes," answered Ravi.

"I am Dimitri Dubov, president of ..."

"I know who you are, Mr. Dubov," interrupted Ravi. "I read an excellent paper by one of your scientists on quantum mechanics in the latest issue of the *Archives of Physics*."

"Well, my scientists are excellent, Ravi, but between you and me, our work has come to a halt," said Mr. Dubov, leaning in close. "That's why I'm here," he continued. "We were very close to making a breakthrough in cold fusion, but we've had a serious security breach. Some of the key work on the secret fusion formula has been stolen. It is worth billions."

"That's very unfortunate, Mr. Dubov," said Ravi with genuine concern.

"From what I hear, Ravi, if anyone can solve this mystery, it's you."

"I'll certainly do my best, Mr. Dubov," answered Ravi. "Please fill me in."

"I have my five top scientists on this project: Dr. Alpers, Dr. Bethe, Dr. Cedrick, Dr. Dobkins, and Dr. Eggland. Every day after work, they lock up the materials and results in a safe. They have devised a security system to prevent theft, but they have not even told me the details. All I know is that the safe has multiple locks on it, and each of them has multiple keys. The system they have devised assures that no one of them, or even any two of them together, could open the safe. However, any three of them are able to open it, so that work can still go on if one or two of them are out at a scientific conference or are sick. They believed that this would keep their work safe, because a majority of them would never collude to steal."

"Have you questioned the scientists, Mr. Dubov?"

"Yes, I have, Ravi. Actually, I questioned them each individually. All of them denied any knowledge of what happened. Of course, I immediately confiscated all the keys from everyone. When I did that, I noticed that Dr. Eggland had one less key than everyone else. He claims to not have noticed that it was missing, because he has not had to use it lately. He does not know who took it, and has not told the others. I do not know if this was an inside job, or if our secrets were stolen by some of our rivals, possibly working with Eggland. I find that more likely, since he is the only one with a missing key."

Ravi listened intently to everything Mr. Dubov had to say. He then turned away, and began to pace up and down his yard, with his eyes half closed. After a few minutes, he looked up and said, "Can you take me to the safe, Mr. Dubov? If you do that, and bring the keys which you have confiscated, I think I can tell you who stole your secret formula."

Ravi was indeed able to solve the crime. How did he do that?

Analysis

With some problems, mathematicians seek to find an answer. With others, they simply try to prove that an answer exists. This is that type of problem. We will not be able to solve this crime with Ravi, but with a careful analysis, we can see how he solved the crime, and how we could solve it ourselves if we were in his position.

The analysis depends on understanding the security system which the scientists devised. For ease, we will refer to the scientists using the first letter of their last names: A, B, C, D, and E. If these scientists devise a system of locks and keys for a safe such that no two of them together can open the safe, but that any three of them could open it, how many locks are needed? How many keys must each scientist have?

How can this information on the arrangement of locks and keys be used to solve the crime?

Solution

We need a system such that any two scientists—(A,B) or (A,C) or (B,D) etc.—cannot get into the safe when they are alone. Therefore, there must be a lock that they cannot open, even when they pool their keys. Therefore, we will assume that each lock is labeled with the initials of the two scientists who cannot open that lock. (Of course, the locks don't have a physical label, but the important thing is to be able to distinguish between them in our discussion.) For example, the lock labeled AB cannot be opened by the pair of scientists (A,B), the lock labeled AC cannot be opened by the pair (A,C), and so on. When we think about the problem this way, we see that there need to be at least as many locks as there are possible pairs of scientists from A, B, C, D and E. Thus, we need

$$\binom{5}{2} = \frac{5!}{2!3!} = 10 \text{ locks.}$$

Now, any pair of scientists who come to the safe will find one lock (the one with their initials on it) that they cannot open.

To figure out how many keys each scientist needs to have, let's look at the situation of one scientist, say scientist E. When he joins any other group of two, such as (A,B), the three of them need to be able to get into the safe. Therefore, he needs to have the key for the lock AB, which the first two cannot open. The situation is similar for any other pair that E might join. Therefore, he needs to have keys for the locks labeled with any pair that can be made from the remaining scientists other than himself, i.e.,

$$\binom{4}{2} = \frac{4!}{2!2!} = 6 \text{ keys.}$$

Therefore, the security system requires ten locks on the safe and requires each scientist to carry six keys. To make the situation more concrete, we can say that in the case of scientist E, he would need to have keys for locks AB, AC, AD, BC, BD, and CD, for a total of six locks. He would not have keys for the remaining four locks. These are precisely the locks which have his initial, i.e., AE, BE, CE, and DE. The situation is analogous for each scientist.

Let us keep analyzing the situation concretely. We can now say that the ten locks would be

AB, AC, AD, AE,
BC, BD, BE,
CD, CE,
DE.

Scientist A would have the following keys: BC, BD, BE, CD, CE, and DE. Meanwhile, scientist B would have keys: AC, AD, AE, CD, CE, and DE. We see that each of these two scientists has three unique keys (A has BC, BD, and BE, while B has AC, AD, and AE). Also, they share three keys in common (CD, CE, and DE). Therefore, when they get together, they can open a total of nine locks (the three unique keys for A, the three unique keys for B, and the three keys that they have in common). However, they cannot open lock AB.

Now, let us look at any other scientist, say E. He has keys AB, AC, AD, BC, BD, and CD. When he joins pair (A,B), he overlaps with A in keys BC, BD, and CD. Meanwhile, he overlaps with B in keys AC, AD, and CD. Thus, together A and B already have five of the six keys that scientist E carries: AC, AD, BC, BD, and CD. However, he brings one unique key to the mix: AB. (See the following table.) This is the one key that the first two do not have, and thus the three together can open the safe. The situation is precisely symmetrical when seen from the point of view of any pair of scientists who are joined by a third.

	AB	AC	AD	AE	BC	BD	BE	CD	CE	DE
A					⊝┯	⊝┯	⊝┯	⊝┯	⊝┯	⊝┯
B		⊝┯	⊝┯	⊝┯				⊝┯	⊝┯	⊝┯
C	⊝┯		⊝┯	⊝┯		⊝┯	⊝┯			⊝┯
D	⊝┯	⊝┯		⊝┯	⊝┯		⊝┯		⊝┯	
E	⊝┯	⊝┯	⊝┯		⊝┯	⊝┯		⊝┯		

Understanding this system, Ravi first labeled each of the ten locks as we did above. He then began trying the confiscated keys from each of the scientists and labeling them based on which locks they opened. He found, of course, that the keys were distributed just as we said above. Using this system, Ravi found that Dr. Eggland was missing key AB from his chain. That key alone could not help a single scientist steal the secret formula, but it could help a pair working together, but only if that pair was (A,B). Thus, Ravi was able to determine that Dr. Alpers and Dr. Bethe had conspired to steal the formula.

This same analysis allows us to solve the same type of problem for any odd number of scientists. Thus, if we have 11 scientists who want to devise a system of locks and keys which keeps any five of them from opening a safe, but lets any six of them do so, they would need

$$\binom{11}{5} = \frac{11!}{5!6!} = 462 \text{ locks.}$$

Meanwhile, each scientist would need to carry $\binom{10}{5} = 252$ different keys. Thus, we can see that while the system is ingenious, it quickly becomes very impractical for a larger number of scientists.

Extension

As we can see from this case, it is very helpful to be able to think logically and in a step-by-step fashion when faced with a difficult problem. That being said, can you figure out the answer to the question on Ravi's logic test? Ravi couldn't read the question, but could read the multiple choice answers:

A)   All of the below.
B)   None of the below.
C)   All of the above.
D)   One of the above.
E)   None of the above.
F)   None of the above.

Sometimes, we look at a problem and we just don't know where to start. In such cases, it is sometimes helpful to just start at the beginning, and think.

If choice A is true, then all answers from B to F are true. But if B is true, then C through F must be false, so this produces a contradiction. Therefore, we can eliminate A. If B is true, then D must not be true; however, "Not one of the above" being true means that we can't select B, and we are faced with another contradiction. Choice C implies that A is true. Since we already showed that A can't be true, C can also be eliminated. Since we have eliminated A, B, and C, we know that D cannot be true. Now, if choice F is true, then A through E are false. However, this means that A through

D are also false, which means E is true. This contradiction eliminates choice F. Therefore, E is the only correct answer. It implies that A, B, C and D are false, which we have established. Also, F is false, because E is true!

Murder at
The Gambit

Ravi was not particularly happy about being in Las Vegas. However, his father had brought the family along to a convention he was attending called "DNA for the DA." Its purpose was to teach district attorneys about DNA testing for their cases. Anyway, Ravi's father had said that after the convention they would drive to Los Angeles to visit the "happiest place on earth." And after visiting Caltech, they might also stop in at Disneyland.

Rather than brave the Las Vegas Strip with his mother and sister, Ravi decided to attend some of the lectures with his father. "May as well learn something," he thought. While the two were at a lecture titled "Follies with Follicles" (about the proper way to collect and preserve hair samples for DNA analysis), Ravi's father was paged. They quietly exited the lecture hall, and Ravi's father went to the pay phones while Ravi watched an elderly woman with a bucket of quarters repeatedly tugging the arm of a slot machine. Ravi's father returned with some urgency in his step.

"What's wrong, Dad?" asked Ravi.

"There's been a murder at a new casino called The Gambit. It was supposed to open tonight. Its owner, Joe "Slick" Bambino had been wanted by the DA's office in Illinois for fraud, but there was never enough evidence to convict him," answered the father.

Ravi was immediately interested. He had been reading the *Las Vegas Sun* over breakfast yesterday, and it had an article on The Gambit's upcoming opening night. Joe Bambino was running a one-time promotional game called "The Odds are in Your Favor." One lucky contestant would be selected by lottery to play the game on opening night. The contestant could put down as much money as he wanted to start, up to $1,000,000, which would be his pot. The game would then be played as follows: one hundred cards are used. Fifty-five say "Win" on their face, while forty-five say "Lose." The cards are then shuffled and placed face down on a table. The player always has to bet half of his current pot. If he picks a "Win" card, he wins the amount of the bet; in other words, he adds half of the pot to his total. If he picks a "Lose" card, then he loses the amount of his bet and he subtracts half of the pot from his total. The player goes through and picks each of the cards one at a time, until he has picked them all. Then, he goes home with his winnings.

Ravi and his father arrived at The Gambit casino. Ravi's father flashed his DA's badge, and they were escorted inside, to a penthouse office. As they entered, Ravi saw a silhouette of a body made with tape on the floor. Apparently, that's where Mr. Bambino's body had been.

One of the officers approached them, ignoring Ravi and speaking to his father, "Thank you for coming. We just wanted to let you know about Slick. We know that your office has been after him for a while."

"Thank you," replied Ravi's father. "Do you know who did this?"

"Not quite," replied the officer. "We lifted three sets of finger prints from the doorknob of his office and have identified them. One belongs to his ex-wife, who apparently was owed a significant amount of back alimony. A second set belongs to Mortie White, the tourist who had been picked to play "The Odds are in Your Favor" tonight. He's an accountant from Tampa, Florida. He had already put $100,000 of his money on deposit at The Gambit to play with. The third belongs to Toby Garcia, who owns the Bull Frog casino next door. The word on the street is that Toby felt pretty threatened by the opening of The Gambit and thought he would lose at least half of his business.

"We've pretty much eliminated Mr. White. When we questioned him, he said he was going to win a bundle of money in tonight's promotional game and would have no reason to kill Slick. The ex-wife says Slick owed her over $600,000. She has no alibi. However, according to her lawyer, she was going to collect the $600,000 with interest. Slick was going to pay an amount equivalent to $600,000 saved at 6% interest per year, compounded quarterly for the ten years he had owed her money. She says it can't be her, because if she killed him, she wouldn't get any of the alimony. On the other hand, we just found out that, in his will, he had left her $1,000,000. She says she didn't know that."

The officer paused for a minute, and then continued, "We have a few men at the Bull Frog talking to Mr. Garcia now. He is our leading suspect, but he seems to have an alibi. The bartender and some customers at the Bull Frog say he was there all night last night. Anyway, I'm sure we'll break that alibi soon enough."

Ravi looked on as the officer was speaking, with a mixture of amusement and concern. He leaned over to his father and said, "Dad, is he serious?"

"Yes, Ravi. Why, what's wrong?"

Ravi looked at his father and said, "Dad, I think the murderer is …"

Who do you think Ravi suspects and why?

Analysis

The solution to this mystery depends on answering the following question: if you put down $100,000 to play "The Odds are in Your Favor," how much money will you have at the end of the game?

Solution

The answer depends on realizing that the order in which you pick the cards makes no difference. For example, if you start with $100 and pick a "Win" card, you have $150. If you then pick a "Lose" card, you end up with $75. However, if you first pick a "Lose," you have $50. If you then pick a "Win," you once again end up with $75. Thus, a "Win"–"Lose" pair, in any order, always reduces your money to 3/4 of what it was. This is because if your pot at any time is Q, a "Win" turns it into $(3/2)Q$; then, a "Lose" turns $(3/2)Q$ into $(1/2)((3/2)Q) = (3/2)((1/2)Q) = (3/4)Q$. Further exploration (or a proof by induction) will show you that the order in which a given set of cards is picked does not affect the end result.

Thus, 55 "Win" and 45 "Lose" cards transform a pot Q into

$$Q\left(\frac{3}{2}\right)^{55}\left(\frac{1}{2}\right)^{45} = Q\left(\frac{3}{2}\right)^{10}\left(\frac{3}{4}\right)^{45}.$$

It is as if there are 45 "Win"–"Lose" pairs followed by 10 "Wins."

So, if we start with $100,000, we end up with

$$\$100,000\cdot\left(\frac{3}{4}\right)^{45}\left(\frac{3}{2}\right)^{10} = \$13.76\,!!$$

Ravi quickly realized that Mr. White would lose nearly all of his money in the "The Odds are in Your Favor" game. Ravi also realized that Mr. White, being an accountant, must have eventually come to the same conclusion, although only after he had put down his life savings to play. Apparently, Mr. White, like almost everyone else who heard the rules, initially thought that he was going to make a lot of money. This misconception was fed by Slick calling it a "promotional game" and instituting a lottery to see who would get picked to play. The only way to avoid losing his money was to kill Slick Bambino. Ravi proceeded to explain this hypothesis to the police. Not being an inherently evil man, Mr. White quickly broke down under questioning and admitted his guilt.

Extension 1

Now that we have gotten the element of surprise out of the way with the unexpected result of "The Odds are in Your Favor" game, let's look a little more deeply at the math.

Once again, let us remind ourselves of the rules. A gambler starts with an amount of money M, playing a certain game. Each time, he bets half of his current pot. If he wins, he gets one dollar for every dollar bet, while if he loses, he loses the money that he bet. Let us say that he plays n times and wins x of these. The money he has at the end of these n turns is given by

$$M(n) = M \left(\frac{3}{2} \right)^x \left(\frac{1}{2} \right)^{n-x}.$$

We will say that the player is losing if he has less money than he started with. The first question we wish to answer is, how many games x must he win to avoid being in a losing position?

To do this, let us calculate the break-even point, where the amount of money he has after n turns equals the starting pot:

$$M\left(\frac{3}{2}\right)^x \left(\frac{1}{2}\right)^{n-x} = M.$$

We can cancel the M on both sides and distribute the powers across the parentheses, recalling the power multiplication rule that $a^b a^c = a^{b+c}$. This gives us

$$\frac{3^x}{2^x} \cdot \frac{1^{n-x}}{2^{n-x}} = 1$$

$$\frac{3^x}{2^n} = 1$$

$$3^x = 2^n.$$

Now, we take the natural logarithm of both sides, recalling the power logarithm rule that $\ln(a^b) = b \ln a$, and we get

$$x \ln 3 = n \ln 2$$

$$x = n\left(\frac{\ln 2}{\ln 3}\right).$$

Thus, we now have a formula for the number of games x which need to be won to break even. If x is less than this, the player will be in a losing position, while if it is greater than this, he will have more money than he started with.

Now, $\frac{\ln 2}{\ln 3} = 0.63093$ to five significant digits. If x is any integer less than $0.63093n$, then the player will be in a losing position. Since we can't play a fraction of a game, we need the integer closest to but less than the real number $0.63093n$. We call this the Floor Function of $0.63093n$, and we'll denote it by $F(n)$. Thus, if $n = 100$, $F(n) = 63$. Therefore, if a player takes 100 turns and wins 63 of them, he will just barely be in a losing position, while if he wins 64 of them, he will have more money than he started with. We can verify this with direct calculations:

$$\left(\frac{3}{2}\right)^{63}\left(\frac{1}{2}\right)^{37} = 0.9029, \text{ i.e., less than 1, while}$$

$$\left(\frac{3}{2}\right)^{64}\left(\frac{1}{2}\right)^{36} = 2.7087.$$

Thus, the odds really were not in favor of Mr. White. To avoid losing money, he would have had to win 64 games, not 55.

Now, we ask a harder question. So far, we have been assigning the number of wins and losses (such as by writing "Win" or "Lose" on cards). What if the gambler is playing a true gambling game, where his odds of winning in any turn are p, how can we calculate, in terms of p, the probability $P(n)$ that after n turns, he will be in a losing position?

The gambler will be in a losing position if he wins any number of games less than or equal to $F(n)$. Therefore, we need to calculate the probability that he wins 0 games, 1 game, 2 games, etc. all the way to $F(n)$ games, and add all of these up. To figure this out, we will rely on the mathematics that we developed in the Solution section of "Basketball Intrigue." If we remember what we deduced there, we see that

$$P(n) = \sum_{x=0}^{F(n)} \binom{n}{x} p^x (1-p)^{n-x}.$$

This is the expression we need to calculate the probability that our gambler will be in a losing position. The probability that he will be in a winning position is just $1 - P(n)$. Finally, we see from the calculation of the break-even point that $\frac{x}{n} = \frac{\ln 2}{\ln 3}$, and this represents the fraction of games won. Therefore, if we assign $p = \frac{\ln 2}{\ln 3}$, the odds of being in a losing position should be very close to 50%.

Extension 2

Slick Bambino's ex-wife might have also had a motive to kill him. If he died, she would inherit $1,000,000. On the other hand, if he lived, she would get an amount equivalent to $600,000 at 6% interest compounded quarterly over ten years. Does the ex-wife make more money if Slick lives or if he dies?

If a certain principal P collects $r\%$ interest for one year, it becomes $P(1 + \frac{r}{100})$. If it collects interest for another year, it becomes

$$\left[P\left(1 + \frac{r}{100}\right)\right]\left(1 + \frac{r}{100}\right) = P\left(1 + \frac{r}{100}\right)^2 ,$$

and for n years it becomes $P(1 + \frac{r}{100})^n$.

If interest is compounded quarterly, then this same process occurs every quarter rather than every year, with the principal collecting $\frac{r}{4}\%$ at every quarter. Over ten years, the money gets compounded 40 times (four quarters per year over ten years). Thus, Mrs. Bambino would make

$$\$600{,}000 \cdot \left(1 + \frac{6}{4 \cdot 100}\right)^{40} = \$1{,}088{,}411.05 .$$

Thus, she makes more money collecting alimony plus interest and would have no motive to kill Slick.

$$D_n = \left[\frac{n!}{e} \right]$$

A Day at the Racetrack

The dinner-time conversation at Ravi's house often revolved around cases that his father was trying, or considering taking to trial, but which had him in some way puzzled. It looked like this was going to be another such dinner. Ravi's father, as whenever he had a case on his mind, was quiet and just picked at his food. Ravi's mother looked at Ravi and smiled. They both knew that it would not be long before the father shared the details to get Ravi's input.

"Ravi," began the father, "let me get your opinion on something."

"Sure, Dad," replied Ravi.

"A group of ten students came to Chief Dobson today complaining that they had been cheated at the racetrack; they wanted Chief Dobson to arrest Tony Lavelle, the owner of the San Simone Racetrack. After hearing their story, Chief Dobson believed them, but does not have sufficient evidence for an arrest. However, he did bring Lavelle in for questioning."

"Give me the details, Dad," said Ravi.

"The students go to Arcadia Middle School. They had been complaining that after school activities were really boring and that they wanted some real excitement. Dr. Ornthall, school principal, suggested an outing to the racetrack to bet on some horse racing. After getting the permission of their parents, they set out in the school van to the track. They decided to bet on a new horse-racing game that Lavelle had come up with for children who wanted to do some betting at the track like their parents. It's called "Rank Your Steed." It is a race that involves just four horses. Before the race, you can buy betting cards, at $10.00 each. You rank the horses according to the order in which you think they are going to finish and then turn in the cards. If you don't get any horses in their right order, then you lose your money. If you get one horse in its right order, you get $10.00 back. If you guess two horses in their right order, you double your money and get $20.00. If you guess three, you get $30.00, and if you get the four horses in their right order, you quadruple your money and get $40.00 back," explained Ravi's father.

"That's an interesting game, Dad," said Ravi. "What happened with the students?"

"Each of them says they filled out a card, guessing at the order of the horses. They decided that they would split their winnings among themselves. Since they had never been to the track before and didn't know anything about the horses that were running, they just decided to each take a guess at ranking the horses. They say each of them took a different guess, so that they would increase their odds of winning. Dr. Ornthall confirms that each of their cards had something different on it. He placed their bets at the window for them. After the race, they went to the racetrack cafeteria for some lunch, and then they came to the window to collect their winnings. Tony Lavelle told them that none of them had won anything, and wouldn't give them any money," answered the father.

"That's ridiculous," said Ravi. "Don't they have some proof of what their bets were?"

"Well, they are supposed to," answered the father. "Each betting card has a carbon copy, which acts as a receipt. The problem is that they lost their carbons. They gave them to Ziggy Price (he's one of the ten students who was on the field trip), and he accidentally left them in the men's room on the sink counter. When they came back to look for them, the carbons were gone. Anyway, none of them could remember their guess. They started arguing about it and got themselves all confused. However, some of them swear they got a horse in the right order. When Lavelle saw that they had no carbons, he said that he had already checked their cards and that none of them had gotten any horses correct. Then, he shredded the betting cards, which he does after every race."

"He said that?" asked Ravi.

"Yes. I have his signed affidavit here from when Chief Dobson brought him in for questioning," relayed Ravi's father. "Chief Dobson says he's crooked, but we just don't have any evidence to arrest him."

"Sure you do, Dad. Tell Chief Dobson to draw up an arrest warrant," said Ravi, wondering what was for dessert.

☉

What is Ravi's evidence? How does he know Lavelle is guilty?

Analysis

All ten students and Dr. Ornthall confirm that each of the ten betting cards had a different ranking for the four horses. Lavelle claims that all ten were completely wrong, without a single card guessing the order of any horse correctly.

To turn this into a problem with which we can work, we'll refer to the horses as horse #1, horse #2, horse #3, and horse #4 based on the order in which they finished the race. In other words, horse #1 is the horse that came in first, while horse #4 is the horse that came in last. We will reference each of the students' guesses based on this labeling of the horses, with the guess specified as an ordering of the horses within set brackets. Therefore, the proper order of finishing results is {1,2,3,4}. A student who guessed this ordering would have guessed for the first position the horse that actually finished first, for the second position the horse that finished second, for the third position the horse that finished third, and for the fourth position the horse that finished fourth, getting all four horses in their correct order. Now, each of the ten students filled in a betting card, guessing one horse to be first, another second, and so forth. Referring to the horses as we have done, each of their guesses corresponds to a permutation of the set {1,2,3,4}.

Of course, it is possible for a guess to have no horses in the correct order; for example, a betting card filled out with {2,3,4,1} corresponds to a guess that horse #2 (the horse that actually ended up finishing second) would come in first, horse #3 (the horse that actually came in third) would be second, and so forth. This particular guess got no horses in their correct position. This is known, in mathematics, as a *derangement*, or as a permutation with *no fixed points*.

If one horse was in the correct position, this would be a permutation with one fixed point.

Lavelle claimed that none of the students got a single horse in their correct position, i.e., that each of the ten betting cards corresponded to a derangement.

Our problem is, for the set {1,2,3,4}, how many derangements are possible?

Solution

This problem, because it involves only a small set, can be solved by brute force. We can list all of the permutations of {1,2,3,4}. Since n elements can be permuted in $n!$ ways, there will be $4! = 24$ such permutations. (Unlike the situation in "Basketball Intrigue," here order does matter!) Then, we can count how many of these are derangements.

However, let us solve the problem more generally, for any set {1,2,...,n}. Let us call the number of derangements D_n.

We will derive a formula for D_n using a *recurrence relation*—a relation that depends on previous results—as follows. If there is a derangement, then horse #1 will not have been assigned his correct position. We begin by looking at the case where horse #1 gets assigned position #2. This can be divided into two subcases:

(a) Horse #2 gets assigned position #1.

(b) Horse #2 does not get assigned position #1.

In case (a), for a derangement to occur, we need the remaining $n - 2$ horses to also get assigned the wrong positions. Since the positions for horse #1 and horse #2 are set, this calculation is the same as if we started with only $n - 2$ horses. Therefore, the total number of derangements in this subcase is simply D_{n-2}.

In case (b), for a derangement to occur, horse #2 cannot get position #1 (that's case (a)), horse #3 cannot get position #3, horse #k cannot get position #k, etc. In this subcase, only one horse's position is predetermined (#1), and the number of derangements is D_{n-1}. The fact that in this instance horse #2 cannot get position #1 rather than position #2 is inconsequential.

We can treat the cases where horse #1 receives position #3, or position #4, or position #k in exactly the same way. Therefore, to account for all possible derangements, there are $n - 1$ such possibilities for all the different incorrect positions that horse #1 can get. So,

$$D_n = (n - 1) (D_{n-1} + D_{n-2}).$$

This gives us a recursion formula by which we can calculate D_4, the number of derangements for $n = 4$.

We know that $D_1 = 0$. In other words, if there is only one element, we cannot get it in the wrong order. Similarly, we know that $D_2 = 1$; i.e., of the two ways to permute {1,2}— {1,2} and {2,1}—one of these is a derangement.

Therefore, $D_3 = 2 (D_2 + D_1) = 2(1 + 0) = 2$.

Building up, $D_4 = 3(D_3 + D_2) = 3(2 + 1) = 9$.

Thus, there are only nine possible derangements (or permutations with no fixed points) for the positions of the four horses. If each of the ten students filled out a card with a different permutation, at least one of them was not a derangement, and Lavelle's sworn affidavit was a lie!

Extension I

Let us investigate two other interesting questions about derangements, using the recursion approach we started in the solution:

1. For a set on n elements, what is the probability, P_n, of a derangement?

2.   We had to build up the derangements one from the other. Can we find a formula, given any n, to make the calculation faster?

The probability of a derangement is again the number of derangements, D_n, divided by the number of possible permutations, $n!$. So,

$$P_n = \frac{D_n}{n!} = \frac{(n-1)(D_{n-1} + D_{n-2})}{n!}$$

$$= (n-1)\left[\frac{1}{n} \cdot \frac{D_{n-1}}{(n-1)!} + \frac{1}{n(n-1)} \cdot \frac{D_{n-2}}{(n-2)!}\right]$$

$$= (n-1)\left[\frac{1}{n} \cdot P_{n-1} + \frac{1}{n(n-1)} \cdot P_{n-2}\right]$$

$$= \frac{(n-1)}{n} \cdot P_{n-1} + \frac{1}{n} \cdot P_{n-2}$$

$$= \left(1 - \frac{1}{n}\right) P_{n-1} + \frac{1}{n} P_{n-2}$$

$$= P_{n-1} - \frac{1}{n}(P_{n-1} - P_{n-2}) .$$

Now, we know that $P_1 = 0$, and $P_2 = 1/(2!) = 1/2$. So, using our formula, we can calculate that

$$P_3 = P_2 - \frac{1}{3}(P_2 - P_1) = \frac{1}{2} - \frac{1}{3} \cdot \frac{1}{2} = \frac{1}{2} - \frac{1}{6} = \frac{1}{3},$$

$$P_4 = P_3 - \frac{1}{4}(P_3 - P_2) = \frac{1}{3} - \frac{1}{4}\left(\frac{1}{3} - \frac{1}{2}\right) = \frac{1}{3} + \frac{1}{4}\left(\frac{1}{6}\right) = \frac{3}{8},$$

and so on. If we look at these calculations in a different form, a pattern emerges:

$$P_3 = \frac{1}{2} - \frac{1}{2 \cdot 3} = \frac{1}{2!} - \frac{1}{3!},$$

$$P_4 = \frac{1}{2} - \frac{1}{2 \cdot 3} - \frac{1}{4}\left(-\frac{1}{2 \cdot 3}\right) = \frac{1}{2} - \frac{1}{2 \cdot 3} + \frac{1}{2 \cdot 3 \cdot 4} = \frac{1}{2!} - \frac{1}{3!} + \frac{1}{4!}.$$

You can experiment with a few more cases to prove to yourself that this pattern continues. For a real challenge—we won't go into the details here—you can prove by induction that the following general statement is true:

$$P_n = P_{n-1} - \frac{1}{n}(P_{n-1} - P_{n-2})$$

$$= \frac{1}{2} + \frac{-1}{2 \cdot 3} + \cdots + \frac{(-1)^{n-1}}{2 \cdot 3 \cdot 4 \cdots (n-1)} + \frac{(-1)^n}{2 \cdot 3 \cdot 4 \cdots (n-1) \cdot n}$$

$$= \frac{1}{2!} - \frac{1}{3!} + \frac{1}{4!} - \cdots + \frac{(-1)^{n-1}}{(n-1)!} + \frac{(-1)^n}{n!}$$

$$= \sum_{k=2}^{n} (-1)^k \frac{1}{k!},$$

for $n = 2, 3, 4, \ldots$. Note that we can start the sum at $k = 2$ because $P_1 = 0$.

Now, we have an equation for P_n that is not recursive! Better yet, from this equation we can get our formula for D_n as well:

$$D_n = n! \sum_{k=2}^{n} (-1)^k \frac{1}{k!},$$

or in expanded form

$$D_n = n!\left(1 - \frac{1}{1!} + \frac{1}{2!} - \frac{1}{3!} + \frac{1}{4!} - \cdots + (-1)^n \frac{1}{n!}\right),$$

since $P_n = D_n/(n!)$.

If you are familiar with the representation of e as an infinite sum, then you may recognize the series inside the brackets (which represents P_n) as approaching $1/e$ as n approaches infinity.

This series, in fact, converges to its limit very quickly, as the table suggests.

n	P_n	n	P_n
2	0.5	6	0.3681
3	0.3333	7	0.3679
4	0.3750	8	0.3679
5	0.3667	9	0.3679

This rapid convergence is because the partial sum varies by only $1/n!$ with each additional addend. Therefore, after $n = 7$, the probability of a derangement remains essentially fixed to four decimal places. Therefore, for large n, the number of derangements can simply be calculated as the closest integer to $n!/e$: D_n = Floor Function of $n!/e$.

Extension 2

From Tony Lavelle's point of view as a businessman, will he (on average) make money on his game, Rank Your Steed?

Again, we'll try to answer the question in general, as if the game were played with n horses and with an initial bet of \$1.00. If the person betting picks a permutation with no fixed points, he gets \$0.00; if he picks a permutation with one fixed point (i.e., guesses one horse in its correct position), he gets \$1.00, etc. Therefore, if he picks a permutation with k fixed points, he gets back k dollars.

Let us call $F_n(k)$ the number of permutations of the set $\{1,...,n\}$ with k fixed points. Therefore, the probability of a permutation with k fixed points is $F_n(k)/n!$. So, Lavelle's expected payout, on average, for a \$1.00 bet is

$$\sum_{k=0}^{n} \frac{kF_n(k)}{n!} = \frac{1}{n!} \sum_{k=0}^{n} kF_n(k).$$

To calculate $\sum_{k=0}^{n} kF_n(k)$, we see that this represents the total of all fixed points in all of the possible permutations of $\{1, \ldots, n\}$. Therefore, let us begin by writing out all of the $n!$ permutations of $\{1, \ldots, n\}$, one on each row, underlining any number in its correct position. Here are some sample rows:

$\underline{1}$	$\underline{2}$	$\underline{3}$	$\underline{4}$	$\underline{5}$	$\underline{6}$	$\underline{7}$	$\cdots$	$\underline{n}$
2	4	$\underline{3}$	1	7	$\underline{6}$	n	$\cdots$	5
3	$\underline{2}$	5	$\underline{4}$	n	$\underline{6}$	$\underline{7}$	$\cdots$	1
$\vdots$	$\vdots$	$\vdots$	$\vdots$	$\vdots$	$\vdots$	$\vdots$	$\cdots$	$\vdots$

So, for example, each time the number 3 appears in the third column, it will be underlined, because it is in the correct position.

In looking at the rows, we see that, for any k, there will be $F_n(k)$ rows with precisely k underlined points. Therefore, the total number of fixed points is for a particular n is $kF_n(k)$, corresponding to all of the underlined points in all of the rows.

Now, let us count all of the underlined points on a column-by-column basis. Each column has $n!$ entries (since there is one row for each permutation), and there are n possible numbers. Therefore, each number appears $n!/n = (n-1)!$ times in each column. Specifically, in the jth column, the number j appears $(n-1)!$ times. Any other numbers in the jth column are not underlined. Therefore, in each column, there are exactly $(n-1)!$ underlined numbers. Since there are n columns, the total number of underlined numbers is $n(n - 1)! = n!$.

Thus, we see that $\sum_{k=0}^{n} kF_n(k) = n!$. Therefore, the expected payout on a \$1.00 bet is

$$\sum_{k=0}^{n} \frac{kF_n(k)}{n!} = \frac{1}{n!} \sum_{k=0}^{n} kF_n(k) = \frac{1}{n!} \cdot n! = \$1.00.$$

So, Lavelle does not make any money on the little game he has set up. That is probably why he felt that he had to steal the money from the students when he had the chance.

We see, of course, that this calculation is the exact same one we would have done to calculate the expected value of k for any random permutation. In other words, if we take any random permutation of $\{1, ..., n\}$, we expect that, on average, one number will be in its proper place, and that this result is independent of n. How about that?

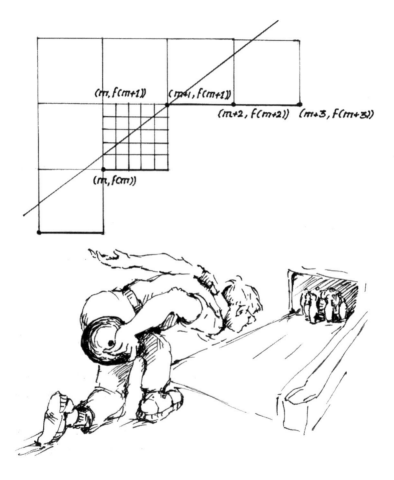

Bowling Average

When Ravi returned home from basketball practice on Tuesday night, he was surprised to see his father sitting in his recliner reading the newspaper.

"What are you doing home, Dad? It's Tuesday night. Shouldn't you be at bowling practice? The league championship is next week," said Ravi.

"I had to give up my spot to Martin Crest, our alternate," replied Ravi's father, crestfallen. He was not good in any sport except bowling, which he truly loved.

"How did that happen, Dad?" asked Ravi, concerned.

The father slowly replied, "We had agreed to keep track of our strike averages. Martin took it very seriously and has been training non-stop for the last two months, even bowling with another league Monday, Wednesday, and Friday. He bought some handheld computer thing that keeps a running track of his average, and he updated it after every single frame. When we first started, he had a 70% strike average. Last night, he called Webster, our team captain, and said that his average was now precisely 90%. Last night, he bowled a 289. Hearing that, Webster gave him the fourth spot on the team instead of me."

"I'm sorry, Dad. How are you taking it?"

"Not at all well," interjected Ravi's mother, who has just walked into the room.

"That's not true," answered Ravi's father, "I'm having just a little trouble believing it—that's all."

"I wouldn't call asking Crest to fax you his average sheet as 'having just a little trouble,'" retorted Ravi's mother.

"Well, I've been working all year toward this championship tournament, and it's a little bothersome to be replaced at the last minute. Anyway, it doesn't matter. Crest's spreadsheet speaks for itself. He raised his strike average from 0.7 to 0.9—that's all there is to it," said Ravi's father, pulling out a stack of papers from under his newspaper and shaking it in the air.

"Well, I'm glad you're not taking this *too* hard," said Ravi sympathetically, giving his dad a little smile.

"I'll be fine. I'm just a little disappointed, but I'll be fine. Ravi, please put these spreadsheets in the garbage for me. I'm going to go wash up for dinner, and let's talk no more about this," said Ravi's father, getting up from the recliner and handing the newspaper and stack of spreadsheets to Ravi.

About ten minutes later, Ravi's father came into the kitchen. Ravi's mother was busy setting the table. Ravi was standing over the kitchen trash can. His foot was on the pedal, and the white lid was open. Ravi was intently studying the spreadsheets. The first number on the first column of the first sheet was 0.70000 and the last number on the last sheet was 0.90000. In between, there were hundreds of numbers representing the running strike average after each frame, such as 0.70149, 0.70297, 0.71154, 0.70813, and so on.

"What are you doing, Ravi?" asked his father.

Ravi looked up, seeming somewhat puzzled. "Dad, where are the rest of the spreadsheets?"

"That's it," replied his father. "What do you see there that's bothering you?"

"It's not what I see—it's what I *don't* see here," replied Ravi. "These are not the spreadsheets from Mr. Crest's computer. These numbers are fabricated, Dad!"

What did Ravi mean?

Analysis

The key here is to figure out what problem we even need to solve. Ravi was looking for a specific value. To understand his thought process, we pose the following problem:

> If a bowler is keeping track of his strike average, or a basketball player is keeping track of his free throw average (or any such sequential binary trial), and the average starts at 70% and improves to 90%, does the average at some point along the way have to equal precisely 80%, regardless of the beginning and ending numerators and denominators which produced 70% and 90%?

If you can solve this problem, you will understand how Ravi proved that Crest had fabricated the spreadsheets. Note that we are dealing with a *binary* trial because, with each attempt (a bowling frame, a free throw), either you reach your goal (a strike, a basket) or you don't.

𝄞

By the way, Ravi's father got his spot back on the team, and they went on to win the league championship.

Solution

If a bowler has a strike average of 0.7, and then he bowls better and increases it to 0.9, does he at some point along the way have to have an average of 0.8?

This is a tricky problem. You might think the answer would surely be "no." It is certainly possible to "skip" a value when taking the average of a collection of values such as the shoe rental prices for all bowling alleys in the United States. But we aren't working with just any values. You may be surprised to find that the answer to our question is always "yes!"

To approach this problem, we have to go back to the term *sequential binary trial*. When we calculate a strike average, what we really have is a sequence of trials (bowling frames or attempts at getting a strike), each of which is either successful or unsuccessful. Let's work with these values as ordered pairs (successful trials, total trials) or (strikes, frames) or, as symbols, (s, f). For example, $(7,10)$ represents ten frames bowled and seven strikes made.

Note that the values s and f must be whole numbers greater than or equal to zero. Also, s must be less than or equal to f. The other interesting fact about our ordered pairs of strikes and frames is that the possible pairs form a *lattice*, as shown in Figure 1.

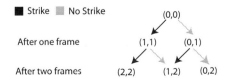

Figure 1. The first steps in forming the lattice.

Let's look at a sample of a larger lattice. The points on the grid in Figure 2 represent all possible ordered pairs for strikes over a total of 20 frames. The path on the lattice shows one bowler's progressing strike average. Over twenty frames, the bowler hit four strikes in a row, then misses twice, next gets three more strikes, then misses twice again, and finally finishes with a hot streak of nine strikes. You can treat specific strike averages as lines that go through the lattice formed by our sequential binary trial. In the figure, we see that the bowler never has an average of 60% or 0.6 (though they come close in frame 11 with 7/11 = 0.63636). In the given lattice, the 60% line passes through four lattice points—(3,5), (6,10), (9,15), and (12,20), marked with gray dots in Figure 2—since

$$\frac{12}{20} = \frac{9}{15} = \frac{6}{10} = \frac{3}{5} = 0.6.$$

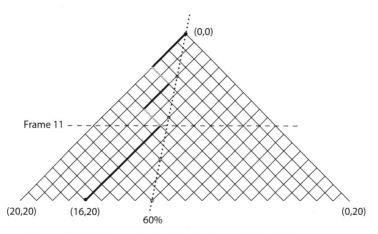

Figure 2. A 20-frame lattice showing a sample bowler's progress and a sample strike average line.

Treat each of the averages 70%, 80%, and 90% as lines over a lattice. The bowler starts on one of the lattice points on the line that corresponds to an average of 70%. He ends up at some lattice point on the line that defines the average 90%. At some point, the path of the bowler's progress must cross the 80% line, which is evident from the diagram in Figure 3(a). This can happen in two ways:

1. The 80% line passes exactly through a lattice point on the path.
2. The 80% line passes through the path in the space between two lattice points.

If the first possibility occurred, then 0.80000 should have occurred on the spreadsheets. Since Ravi could not find 0.80000 anywhere, let's assume that the second possibility occurred for Crest's average to improve from 70% to 90%. Therefore, the 80% value must sit between two successive spreadsheet entries; that is, the 80% line crosses in between two successive lattice points. Let's call the points (s_k, f_k) and (s_{k+1}, f_{k+1}), where

$$\frac{s_k}{f_k} < 0.8 < \frac{s_{k+1}}{f_{k+1}},$$

or, in easier-to-use fractions,

$$\frac{s_k}{f_k} < \frac{4}{5} < \frac{s_{k+1}}{f_{k+1}}.$$

Because we are working with lattice points, $f_{k+1} = f_k + 1$. If the bowler does not get a strike in frame f_{k+1}, then $s_{k+1} = s_k + 0 = s_k$. However,

$$\frac{s_k}{f_k + 1} < \frac{s_k}{f_k}.$$

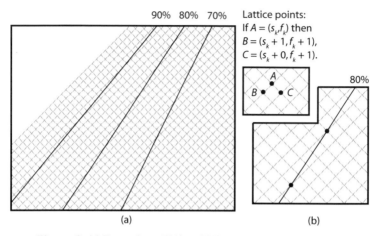

Figure 3. (a) To get from 70% to 90%, you must pass over 80%. (b) Visual "proof" that the bowler's path must land exactly on the 80% line.

Therefore, it must be the case that the bowler gets a strike in frame f_{k+1}, in which case $s_{k+1} = s_k + 1$. This gives us

$$\frac{s_k}{f_k} < \frac{4}{5} < \frac{s_k + 1}{f_k + 1}.$$

The expression can be broken up into two inequalities:

$$\frac{s_k}{f_k} < \frac{4}{5} \text{ and } \frac{4}{5} < \frac{s_k + 1}{f_k + 1}.$$

The first is the same as $5s_k < 4f_k$, and the second gives us $4f_k + 4 < 5s_k + 5$ or $4f_k < 5s_k + 1$. Combined we have

$$5s_k < 4f_k < 5s_k + 1,$$

but this cannot be true. There does not exist an integer $4f_k$ that lies between the integer $5s_k$ and the next positive integer $5s_k + 1$. That would be like saying four times some integer is in between 20 and 21—it can't happen.

In other words, if the strike average passes over the 80% average line, it must be the case that it lands exactly on the 80% line. Use Figure 3(b) to verify this fact visually.

Because 0.80000 did not appear on any of Crest's spreadsheets, Ravi knew that the numbers had to be fabricated.

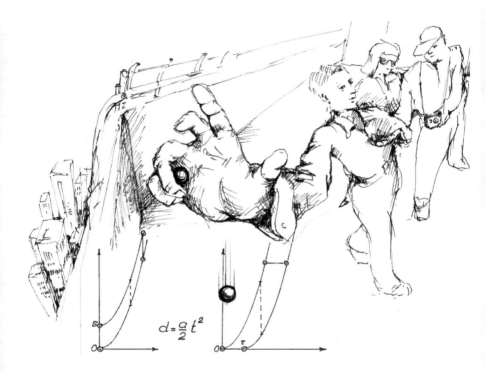

$$d = \frac{a}{2}t^2$$

Caught on Film

When the day began, it seemed like a typical lazy Saturday morning. Ravi and his parents chatted while they ate breakfast. Then, the phone rang. Ravi's mother had come to dread the phone ringing on a Saturday morning, as it often meant that her husband, a Chicago district attorney, would be called in on a case. And so it was. Ravi's mother answered the phone and, rolling her eyes, handed the phone to her husband, saying, "It's Chief Dobson. He says he needs you on a case."

After Ravi's father hung up, he turned to Ravi and said, "I have to go to the Sears Tower. Apparently, a couple of teenagers dropped two steel marbles from the top of the Tower. One of the marbles hit someone at the bottom. He is now in critical condition at Mercy Hospital. Do you want to come with me?"

"No thanks, Dad. I was going to finish working on a computer program this morning," replied Ravi.

"Come on, forget about that this morning. Come keep me company, and we'll stop by and get you those doughnuts you love on the way," coaxed his father.

This was a hard offer to turn down. The only thing Ravi liked more than doughnuts was math, although his lean frame belied this affinity for doughnuts.

"Alright, Dad, you talked me into it," said Ravi smiling.

When they arrived at the Sears Tower, the area had been cordoned off and there were several police cars by the main entrance. As they walked through several uniformed police officers, Ravi was busy brushing powdered sugar off his blue shirt. Chief Dobson met them and quickly explained the situation. "Luckily," he said, "it was all caught on film. The victim was an actor who was in the middle of shooting a scene at the base of the Sears Tower for an insurance commercial. Then, the steel marble hit him on the head, knocking him unconscious."

"I thought there were two marbles," interjected Ravi's father.

"Yes, there were," answered the Chief. "The second one landed after the first and just off to the side. Looking at the film, it was about 30 feet off the ground when the first one hit. It took a bit of looking to find that second one in the air, but luckily the crew was shooting with a wide angle lens."

"Who are the perpetrators?" asked Ravi's father.

"Two teenagers," gruffed Chief Dobson. "We have them and the witnesses still up on the roof."

After a long elevator ride, Ravi, his dad, and the Chief stepped out onto the outdoor observation deck at the top of the tower, at a height of 1431 feet. A uniformed police officer came over and immediately summarized what they had discovered so far. Apparently, the suspects were a sixteen-year-old named Joseph Hendrix and a fifteen-year-old named Tommy Aston. A group of four tourists signed statements that they saw Joseph extend his hand quickly over the railing and then a split second later, Tommy extended his hand over the railing. Both boys looked over the edge and then suddenly turned and started running for the elevator. Luckily, there was a large group already on the down elevator, and the elevator operator would not let them on, forcing them to wait for the next one. In the meanwhile, security radioed from downstairs saying that someone had thrown a projectile over the railing, and the elevators were shut down, trapping the boys on the deck.

"Did the witnesses see the marbles actually being dropped?" Ravi asked the police officer.

"No. No one actually saw the marbles fall, because it all happened so quickly. But they know where the boys were standing, and the marbles landed directly below that area," answered the officer.

"That's enough for me," said the Chief. "Let's go get a confession out of these punks," the Chief said to Ravi's father.

Chief Dobson and Ravi's father questioned each of the two suspects in turn while Ravi looked on. Ravi knew that it was standard police practice to question the suspects separately. It gave the police a chance to find inconsistencies in their stories and gave each suspect a chance to implicate the other.

Joseph Hendrix, a tough looking, spiky-haired kid, refused to answer any questions, continually glaring at Chief Dobson and saying, "I'm a juvi and I ain't sayin nothin'."

Tommy Aston, by contrast, looked pale and frightened, perhaps beginning to appreciate the seriousness of his situation. He promised to tell the police everything. He proceeded to state that Joseph had thrown both marbles over the railing and that he himself had nothing to do with it.

"We have witnesses that say they saw you put your hand out over the railing just after Joseph. We know that you threw the second marble. There's no use denying it, and I promise you things will go much easier on you if you confess," Ravi's father said in an angry tone.

"No, no, you've got it all wrong!" protested Tommy. "I put my hand out to try to stop him, as soon as I realized what he was going to do."

As Ravi watched Tommy squirm before his father and Chief Dobson, he began to feel sorry for him.

"So you're claiming Hendrix threw *both* marbles, young man?" questioned Chief Dobson in an incredulous manner. "He had his hand over the railing for only an instant. All four witnesses said so. He did not stick his hand out a second time. But *you* stuck your hand out after him."

"I told you, I was trying to stop him," pleaded Tommy, now on the verge of tears.

"So who threw the second marble, then?" continued Chief Dobson relentlessly.

"He threw them both at the same time," answered Tommy.

"Rubbish!" retorted the Chief. "We've got it all on film. The second marble was 30 feet above the ground when the first one hit. It would have been impossible for him to have thrown them both together."

The Chief turned to one of the uniformed police officers. "We're done here. Take them downtown," he said, gesturing to Tommy and Joseph.

"Wait a minute, Chief," interjected Ravi. The police officer stopped. Instinctively, he had come to consider Ravi a person of authority because of the high regard that the Chief had for him, and because of his uncanny ability to solve cases that baffled everyone else.

"What is it, Ravi?" asked the Chief. He knew from experience that if Ravi had a concern, it should be addressed.

"There's something I'd like to check out," said Ravi. He walked over to the Chief, and whispered to him, "I'm not so sure that this kid is lying. I'd just like to check something in his story. I will need two steel marbles like the ones tossed over the edge, as well as the witnesses. Also, do you think the camera crew would be willing to help? I'll need them to shoot some footage of me dropping the marbles, to see if we can get the timing of what happened," stated Ravi in a distracted manner, already thinking ahead to the various scenarios he wanted to test.

"I'm sure the crew would do anything to help, Ravi. But we can't be long. We've been holding the witnesses here all morning. Also, as much as I respect your mind, I've had enough experience with these juvenile delinquents to know when they're lying," replied the Chief.

Soon, Ravi was standing before the witnesses, with the camera crew behind him, rolling film.

"Um... excuse me, everyone," began Ravi slowly, addressing the witnesses. "I need to get from you a sense of how quickly the first suspect put out and withdrew his hand. I'll try a couple of things, and you tell me which matches what you saw," said Ravi, confidently addressing the four adults standing impatiently before him.

Ravi put both marbles in his right hand, one next to the other inside his closed fist, which he held horizontally. He extended his hand and quickly unfurled his thumb, index and middle fingers, letting the first marble drop. It hit the ground in less than a second, and he then let the second marble drop by unfurling his ring and fifth fingers, quickly withdrawing his hand. The entire maneuver took less than two seconds.

"Is this how fast Hendrix put out and withdrew his hand, or was that too fast?" asked Ravi, turning to the witnesses.

The witnesses looked at each other, and each of them said in turn that this was actually too slow, not too fast. Ravi then repeated the maneuver several more times, faster and faster each time. On the last trial, he quickly extended his hand, and unfurling his fingers, let the marbles drop directly one after the other, immediately withdrawing his hand back.

The witnesses all concurred that this last trial was most similar to the speed at which Hendrix's hand had moved.

"Thank you all very much," said Ravi. "I'm done."

After the witnesses were dismissed, Ravi sat with his father and the Chief to review the footage. They scrutinized the last trail very carefully, stepping through the footage frame by frame. As Ravi unfurled his thumb, index finger, and middle finger, the first marble fell from his hand. He continued unfurling his fingers in one smooth motion, and the second marble was released almost immediately thereafter. Using the calibrated measuring tool on the film viewer, the first marble had fallen only about 2 inches when the second one was released.

"That does it Ravi," said the Chief. "As you can plainly see, when you moved as fast as Hendrix, the separation between the marbles was only 2 inches, not 30 feet. The marbles will obviously fall at the same speed. Therefore, Hendrix could not have dropped both marbles. Hendrix dropped one and Aston dropped the other. It is the only logical explanation."

"I completely concur, Chief. I will charge them both," said Ravi's father. Both men turned and began walking toward the elevator.

Abruptly, however, they were stopped by Ravi's voice: "Hold on there, gentlemen. Let's get this right, shall we?"

Analysis

Before we begin looking at this problem, we need to know a little something about the physics of falling objects. We'll keep this simple by neglecting air resistance, since the objects we are discussing are small steel marbles, which should have only minimal air resistance.

Falling objects are basically bodies moving under the acceleration of gravity, so the equations we need to understand their behavior are the same as those for any objects moving under a constant acceleration.

For any object moving at a constant velocity (speed), we know that distance is simply the product of velocity and time: $d = vt$, where d is distance, v is velocity, and t is time.

However, objects moving with a constant acceleration are constantly changing their speed. For example, if an object starts with a velocity of 5 feet per second but accelerates at 5 feet per second per second (5 ft/sec^2), then after 1 second, its velocity is 10 ft/sec, after 2 seconds, its speed is 15 ft/sec, etc. The equation for velocity as a function of time under those conditions is: $v_f = v_0 + at$, where v_0 is the intial velocity, v_f is final velocity, and a is acceleration. Now, velocity is not a constant but is a function of time. This function is linear when acceleration is constant.

This makes the calculation of distance as a function of time more difficult. One fact that always holds true with changing velocity, however, is that distance will remain the product of the *average* velocity and time: $d = v_{avg}t$. Now, if the velocity is increasing linearly with time, the average velocity is simply $(v_0 + v_f)/2$, where v_0 is the starting velocity and v_f is the final velocity due to the constant acceleration. Therefore, we can say that

$$d = \left(\frac{v_0 + v_f}{2} \right) t.$$

Now, we can substitute $v_f = v_0 + at$ into this equation, and we get,

$$d = v_0 t + \frac{1}{2} at^2.$$

If the starting velocity of the body is zero, we have

$$d = \frac{1}{2} at^2.$$

This is going to be our key equation for a falling body starting at rest (i.e., simply released from someone's hand). In this case, the acceleration, a, is the acceleration due to gravity, which is known to be 32 ft/sec². You may at this point be thinking that this is very unfair, because if you haven't had physics, how are you supposed to know this value that you need to solve the problem? We'll get back to this later!

Now, we are ready to catch up with Ravi and look at this case. The Chief and Ravi's father made the following argument (without the equations): both steel marbles fall identically, so if they begin 2 inches apart, they will remain 2 inches apart all the way down. This seems to make sense, since the fall of both marbles is governed by the equation

$$d = \frac{1}{2} at^2.$$

What is wrong with this reasoning? Can you see the mistake the Chief and Ravi's father made?

The mistake is that they posed the problem incorrectly. They were analyzing the following problem: if two identical steel marbles are positioned 2 inches apart, one above the other, and dropped at the same time, how high will the second one be when the first hits the ground? However, if we think carefully about the situation, we see that the correct problem is as follows: a steel marble is dropped and falls 2 inches. At that instant, a second steel marble is dropped from the same height as the first, and they both continue to fall. How high is the second marble when the first one hits the ground?

Does it make a difference when the problem is stated in this way? This is now your problem to solve.

Solution

To make the solution a bit easier, we'll assume that the marbles fall from the top of the Sears Tower to the ground (actually, the first one hit the victim's head, about six feet off the ground, but that won't make a big difference to our argument).

So, how do we solve this problem? First of all, let us decide the units in which we will work: feet and seconds, with the acceleration due to gravity equaling 32 ft/sec². Now, let us call the height of the Sears Tower h (1431 feet to the highest occupied floor). Now, how about the following approach: let us calculate the time t_1 it takes for the first marble to fall $h - 0.167$ ft (i.e., 2 inches less than the height of the Tower). We can do this using the equation

$$h - 0.167 = \frac{1}{2} a t_1^2.$$

Then, we can use t_1 to calculate the distance that the second marble falls. Sounds good, except for one problem: it's completely wrong! We've again fallen into the trap of assuming that the first marble starts from rest, but two inches ahead of the second marble. That is essentially the same mistake the Chief and Ravi's father were making. So how then do we solve the problem correctly?

I remember a line from a movie where a gangster is threatening someone to get information out of him, and he says, "We can do this easy, or we can do it *real* easy. Your choice." Let's do it real easy.

Let us call h the height of the Sears Tower and t the time it takes an object (with no air resistance) to fall height h. Let us call s the distance that the first marble falls before the second marble is released and τ the time it takes the first marble to fall distance s. Our question then becomes, very simply, how far does the second marble fall in time $(t - \tau)$? We'll call this distance h'. Thus, we have the following equations:

$$h = \frac{1}{2}at^2 ,\ s = \frac{1}{2}a\tau^2 ,\ h' = \frac{1}{2}a(t-\tau)^2 .$$

What we are seeking to find is how high the second marble is when the first one hits the ground, in other words, $h - h'$. We calculate this difference:

$$\begin{aligned}
h - h' &= \frac{1}{2}at^2 - \frac{1}{2}a(t-\tau)^2 \\
&= \frac{1}{2}a(t^2 - (t^2 - 2t\tau + \tau^2)) \\
&= at\tau - \frac{1}{2}a\tau^2 \\
&= at\tau - s.
\end{aligned}$$

Rearranging the first two equations, we have

$$t = \sqrt{\frac{2h}{a}},$$

$$\tau = \sqrt{\frac{2s}{a}}.$$

Now, we can substitute these quantities into our formula for $h - h'$ and get

$$h - h' = a\left(\sqrt{\frac{2h}{a}} \cdot \sqrt{\frac{2s}{a}}\right) - s = 2\sqrt{hs} - s .$$

In our problem, h is the height of the Sears Tower, about 1431 feet, and s is 0.167 feet (i.e., 2 inches). Therefore, $h - h'$ works out to be 30.75 feet.

This is a pretty surprising result, that if one marble gets just a 2-inch head start on the other, by the time they have fallen the height of the Sears Tower, the distance between them has increased to over 30 feet. No one had expected this, but when Ravi laid out the calculation, it demonstrated that the evidence was very consistent with the notion that Hendrix had actually dropped both balls. Ultimately, only he was charged with the crime, and Aston's testimony was used to convict him.

Finally, there is one loose end. For those who thought the problem was unfair because it seemed as if you needed to know the value of the acceleration of gravity ($a = 32$ ft/sec^2) to solve it, you see that, in the solution we found, we actually never needed to know this value at all—it cancelled itself out in the equations, and we never used it once!

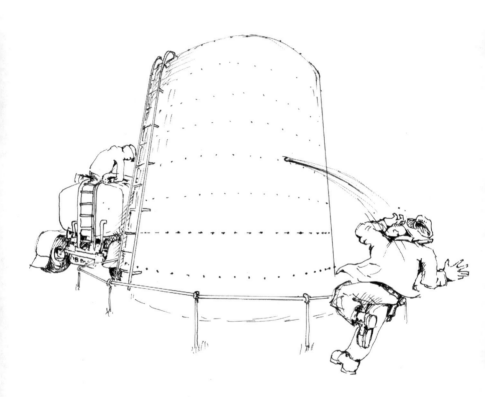

A Mishap at
Shankar Chemicals

The lamb vindaloo was especially tasty this evening. Ravi's mother had outdone herself. Yet, Dr. Sanjiv Lavishankar only picked at it with his fork. He clearly had no appetite. The events of last week had left him quite saddened. This is why Ravi's father had invited his friend over for dinner that evening—in an attempt to cheer him up. Yet, Dr. Lavishankar could not forget his troubles and recounted his story once again to the family as if still trying to come to terms with the fact that it had actually happened.

Dr. Lavishankar was president and founder of Shankar Chemicals, a successful chemical plant located in a wooded area in southern Wisconsin about an hour north of Chicago. Because of Ravi's interest in chemistry, his father had taken him to visit the plant, and Dr. Lavishankar himself had served as their tour guide. Ravi saw the process of isolating and purifying various chemical compounds. The most visually startling sight at the plant was the collection of large, metallic, cylindrical tanks which housed the many chemical solutions that the plant manufactured and stored for sale to various industries.

A few days ago, an unexpected tragedy had occurred at the plant. Joseph Stackhouse, a security guard for Shankar Chemicals, was on his patrol one afternoon when he was doused by a chemical from one of the vats as a result of a freak accident. As best as anyone could figure, a hunter in the nearby wooded area must have shot a stray bullet that punctured one of the tanks. As the tank was punctured and liquid began to squirt out, Joseph Stackhouse was doused by the chemical solution before he could get out of the way. Luckily, the tank that sprayed Mr. Stackhouse contained a 40% solution of acetic acid, a relatively weak acid which causes only mild burns to the skin.

Dr. Lavishankar was relieved when he first visited Joseph Stackhouse and found his injuries fairly limited. The doctors treating Stackhouse said that he would make a full recovery within a matter of weeks, with the only lasting damage expected to be some minor depigmentation on his arms.

Yesterday, though, Dr. Lavishankar received a call from Mr. Stackhouse's attorney, informing him that Mr. Stackhouse intended to bring immediate suit against Shankar Chemicals for the physical and emotional injuries sustained by Stackhouse. The attorney stated that since there are sometimes hunters in the area, Shankar Chemicals should have fortified its tanks to guard against bullets.

Dr. Lavishankar tried to explain that hunters never strayed near the grounds as there was no game in the vicinity of the plant. Also, the likelihood of such an accident occurring was infinitesimal, he explained. Nevertheless, Dr. Lavishankar clearly understood that this lawsuit could financially ruin the business he had spent the better part of his life building. Moreover, he felt quite guilty about what happened to Stackhouse and was thinking of not even contesting the suit. As Ravi listened, he also thought that the odds against such an occurrence were astronomical.

The more Dr. Lavishankar talked, the sadder he became. Finally, he stopped talking and pulled folded papers from his jacket pocket and tossed them on the dinner table, "It's all here—the story of the end of Shankar Chemicals."

Ravi felt sorry for his father's friend, for he knew what a dedicated scientist and decent person he was. Several minutes of awkward silence passed, making everyone at the table self-conscious. Just to do something, Ravi reached for the folded up incident report that Dr. Lavishankar had tossed onto the table and started flipping through it.

The first page of the report indicated that Mr. Stackhouse was found by the driver of a supertanker truck, who had just started pumping liquid from the other side of the same vat that had been punctured. The driver called an ambulance, and because he had heard a gunshot, he called the police as well. As Ravi looked through the incident report, he thought to himself that the police had done a rather thorough job with the investigation, noting all the details of the incident. For example, the vat was a cylinder 20 meters high and 10 meters wide. Ten meters out from the vat, a rope cordoned off a safety zone so that visitors to the plant would not approach too close to the tank. The driver of the tanker truck had logged in his arrival to the plant at 4:12 p.m. He checked the gauge on the vat and noted in his log that the tank was full to capacity. He had hooked his truck up to the vat and began pumping liquid from the tank at 4:26 p.m. After about 25 minutes, the driver recounts that he heard a shot. He immediately shut the pump down, and the tank clock noted that its valve closed at 4:52 p.m. The tanker truck driver reported that about a minute after the shot, Stackhouse came running from the other side screaming, "Help me! Help me! I've been sprayed!"

It was at this point that the truck driver called the ambulance and then the police. When the police arrived, they indeed noted liquid trickling out of a small bullet hole in the tank—they measured the hole at 9.5 meters off the ground.

Ravi was quite impressed that the police had even gone to the hospital and took a statement from Stackhouse once it became clear that his injuries were mild. Stackhouse could not provide a precise time for the gunshot, saying it was between 4:30 and 5:00 p.m. He stated that he was walking the perimeter of the tank, just inside the roped off area, as he usually did. He knew that a tanker had come in to pump some chemicals for delivery, but he could not see the truck as it was on the other side when the shot rang out. As he heard the shot, he felt the liquid chemical spray him before he could move out of the way. Realizing what had happened, he ran around the tank to the driver and yelled for help.

By this time, Ravi was so engrossed in the details of the report that he had not noticed that everyone else had left the dining room table and gone into the living room. Ravi slowly eased his chair back from the table, but remained seated for an additional few seconds, still engrossed in thought. He then walked in the living room and interrupted the conversation taking place, "How fast does the truck pump liquid from the vat?"

Everyone looked up at Ravi, surprised. His mother immediately felt embarrassed by Ravi's abruptness, for she had taught him not interrupt without saying, "Excuse me." Indeed, it was unusual for Ravi to forget his manners as he was quite a polite young man. Dr. Lavishankar responded, "You would be amazed at the speed, Ravi. Our new high pressure pumping system actually pumps at 3000 liters per minute."

Ravi thought for another few seconds, closing his eyes as he did when visualizing a mathematical equation. When Ravi opened his eyes, he asked Dr. Lavishankar, "Has the bullet been recovered from the tank?"

"No, it is presumably at the bottom of the tank," answered Dr. Lavishankar.

"Well, I think you're going to need that bullet," replied Ravi.

"Why?" Ravi's father interjected.

"To prove that Stackhouse is a liar," answered Ravi confidently.

Analysis

The entire problem about which Ravi thought is the following: if the tank was pierced by a bullet, putting a hole in it, how far would liquid squirt out of the tank at the instant the hole was made and the liquid first started to flow? In thinking about this, he found a glaring discrepancy in the story told by Joseph Stackhouse.

It is probably easier to think about this problem by putting it on more familiar turf. Instead of thinking of a large tank at Shankar Chemicals filled with acetic acid, let us consider a simple tin can filled with water, and ask the same question.

Hint: This problem needs a touch of physics, though I am assuming that most of the readers of this book have probably taken high-school physics. Here is a quick review.

First, think about what makes the water squirt out of the can. Where does the energy to propel the water from the hole in the can come from?

The principle underlying the solution is that of the *conservation of energy* in a closed system. The kinetic energy of the water squirting out of the hole is equal to the *change* in the potential energy of the water in the can.

Recall that the general formula for kinetic energy of a given mass, M, traveling at a given velocity v, is $1/2\ Mv^2$, while the potential energy for a given mass, M, at a height h, is Mgh, where g represents the acceleration of gravity.

Using only this information, we can say that if an object is dropped from a height h, its velocity as it hits the ground, v_f, can be calculated by assuming that all of its potential energy is transferred to kinetic energy. Therefore, if we start with

$$\frac{1}{2} M v_f^{\,2} = Mgh \,,$$

then we immediately get that $v_f = \sqrt{2gh}$.

Now, we know that the initial velocity of the object is zero, and we have calculated its final velocity as its hits the ground. We also recall, as discussed in the solution to "Caught on Film," that the velocity of a falling object increases linearly with time as it falls due to the acceleration of gravity and that the time it takes to fall is

$$t = \sqrt{\frac{2h}{g}}.$$

This gives us all the tools we need to solve our problem.

Be careful, now, since we are going to reuse some of the above variable names in the solution, but they will stand for different things. The important issue is to understand the physical concepts and to be able to apply them.

Solution

Consider this problem:

> Given a cylindrical can of height h filled with water, if
> we put a hole in the can, water spouts out and trav-
> els a certain horizontal distance a before hitting the
> ground. We ask the following questions:
>
> Does this horizontal distance a depend on d, the dis-
> tance from the top of the can to the hole? Is it possible
> to predict how far the jet of water will travel when
> the water first starts to flow? In other words, can a
> theoretical model be constructed for this problem to
> predict a as a function of d?

This is a difficult question, as there seem to be compet-
ing influences on the distance that the water travels. The
water will be forced through the hole possibly based on the
pressure exerted by the water above it. Therefore, the lower
the hole (i.e., the bigger d), the more pressure there is forcing the water out. The water will squirt out of the hole with
a certain horizontal velocity v and fall a vertical height

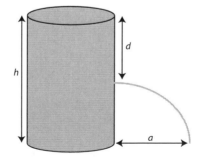

Figure I. A tin can full of water with a hole in its side.

$h - d$ to the ground. The time t it takes the water to hit the ground will depend only on this height. The horizontal distance the water travels will be a function of the horizontal velocity v and the time t it takes the water to fall: $a = vt$.

Presumably, the bigger the value of d (i.e., the lower the hole), the greater a will be because of the greater value of the horizontal squirt velocity v. However, there is correspondingly less time, t, for the water to travel horizontally, because the fall distance, $h - d$, is less.

In order to find the horizontal squirt velocity, we want to calculate the kinetic energy of the water that leaves the can when the hole is made. Due to the laws of conservation of energy, that will be the same as the change in the potential energy of the water in the can. Consider the diagram in Figure 2(a), where we have split the volume of the can into three sections. Let us call them Region 1, Region 2, and Region 3. Each region has a set potential energy E_1, E_2, and E_3, respectively, before a hole is made in the can; the total energy of the water in the can is $E_1 + E_2 + E_3$.

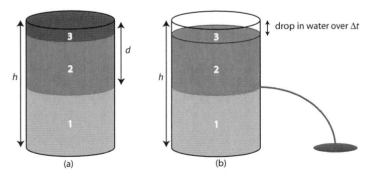

Figure 2. The water in the can (a) before the hole is made and (b) at time Δt after the hole is made.

Now consider the situation in Figure 2(b), some small amount of time, Δt, after the hole is made. Also recall the factors that affect potential energy: the mass and the height off the ground of that mass. Region 1 contains the water below the hole, which is not changed by the existence of that hole. Therefore, the potential energy of Region 1 is still E_1. Regions 2 and 3 are above the hole; since water is leaving the can through the hole, the energy of one or both regions should be altered. When we look at the potential energy of the water in Region 2, we see that it has not changed. It is not the same water molecules (we think of the water draining out the bottom of Region 2), but the water that shifted into Region 2 has the same energy as the water that was there before the hole: the same mass resting at the same height. So, the total energy of the water in the can in Figure 2(b) is $E_1 + E_2$; Region 3 is empty, so it has zero mass and zero energy!

To find the value of E_3, let us start by finding its volume. The base of the can has a given area; call it A_c. As water leaves the can, the water level in the can drops at a given speed; call it v_c. Over the small change in time Δt, the water level has changed by $v_c \Delta t$. So, the volume of Region 3 is $A_c v_c \Delta t$. The mass of the water is that value multiplied by the density of water, which we will call W. That mass of water used to be at the top of the can; after the hole is made, an identical mass left through the hole. The difference in height between those two states is d. Therefore, the change in potential energy inside the can is

$$WA_c v_c \Delta t g d.$$

The mass $WA_c v_c \Delta t$ is the same as the mass of the water squirting out the hole. So, we can write the kinetic energy of the water leaving the can as

$$\frac{1}{2}WA_c v_c \Delta t v^2.$$

Therefore, the conservation of energy equation can be written as

$$\frac{1}{2}WA_c v_c \Delta t v^2 = WA_c v_c \Delta t g d.$$

Canceling terms, we get

$$\frac{1}{2}v^2 = gd.$$

Therefore, the horizontal speed of the water as it exits the hole is $v = \sqrt{2gd}$. It is pretty amazing that this is the same as the vertical speed of an object as it hits the ground when dropped from a height of d!

The distance that the water has to fall vertically is $h - d$. Therefore, the time it takes to do this is

$$t_{fall} = \sqrt{\frac{2(h-d)}{g}}.$$

Now we have the information need to write an equation for a that depends on the variable d:

$$a(d) = v_h t_{fall}$$
$$= \sqrt{2gd}\sqrt{\frac{2(h-d)}{g}}$$
$$= 2\sqrt{d(h-d)}.$$

This equation can be used to predict a as a function of d for a can of given height h.

The expression under the square root sign in the equation $a(d) = 2\sqrt{d(h-d)}$ is the geometric mean of the two values d and $h - d$, which add up to h. The maximum value that this geometric mean can have is the arithmetic mean of d and $h - d$, or $h/2$. (Note that this maximum would occur only when $d = h - d$, or $d = h/2$.)

Therefore, the maximum distance a is reached when $d = h/2$, or the hole is in the middle of the can. This maximum distance is

$$a_{max} = 2\max\left(\sqrt{d(h-d)}\right)$$
$$= 2(h/2)$$
$$= h.$$

In other words, the maximum distance is the height of the can!

Amazingly, both $a(d)$ and a_{max} do not depend on the diameter of the can or the size of the hole!

♀

Now, our problem is solved. All Ravi had to do was figure out how far the liquid could squirt from the tank. From the story, we know that the tank is 20 meters high and 10 meters in diameter. It was full when the giant tanker began to pump some of the liquid out. The supertanker pumps liquid out at 3000 liters per minute, and it pumped for 26 minutes, taking 78,000 liters out of the tank. The radius of the cylindrical tank is 5 meters, therefore, its base area is $\pi(5m)^2$, or approximately 78.5 m². Recalling that a volume of 1 ml is contained in 1 cm³, we see that 1 m³ is equivalent to $(100 \text{ cm})^3$, or 10^6 ml. This is equivalent to 1000 liters. Thus, for a cross-sectional area of 78.5 m², a drop in the liquid level of 1 m is equivalent to a loss of about 78,500 liters.

Thus, at the time the bullet was fired, the level of liquid in the tank had dropped from 20 meters to just over 19 meters. For the sake of argument, we can consider the level to be 19 meters. This would be equivalent to a can with h = 19 m. (It doesn't matter how tall the can actually is, just the level of liquid above the ground. Thus, if the liquid is 19 m high, we can consider that the can is 19 m high). The bullet hole was 9.5 m above the ground. In other words, it was also 9.5 m below the level of the liquid, equivalent to d = 9.5 m in our equations. These are precisely the conditions needed to attain maximum squirt distance—the hole is in the middle of the can! Thus, the liquid would squirt a distance equal to h, or, in this case, 19 m before hitting the ground.

Yet, Joe Stackhouse testified that he had been walking within the roped perimeter 10 meters away from the tank when the shot rang out and that he was doused before he could move away. This, of course, is physically impossible. (The average person is less than 2 m tall.)

The only conclusion is that he had fabricated the story and made the hole in the tank himself, using his own gun. After dousing himself with the acetic acid (which at 40% concentration produces only minor burns), he ran to the driver of the supertanker and claimed he had been sprayed. This was all in an attempt to win significant financial damages from Shankar Chemicals.

When Ravi presented this evidence, a warrant was issued for the arrest of Joe Stackhouse. The tank was drained, and the bullet recovered. Ballistics matched it to the gun that Joe Stackhouse carried while on the job.

Almost Expelled

The classroom door opened abruptly, and without so much as an "excuse me," Mr. Danzing walked in, followed by two eleventh-grade students, Allen Kaitelle and Bill Hennings. "Can I help you, Mr. Danzing?" asked Mr. Shelby, the American history teacher, somewhat surprised and amazed at this interruption of his third-period history class, especially when he was reviewing key material for next week's final exam.

Mr. Danzing, the precalculus mathematics teacher, realized belatedly his rudeness in bursting in on Mr. Shelby's class. Mr. Danzing was a fine and dedicated teacher, but was known to all his students as rather excitable—and when he got excited, he became impulsive.

"I sincerely apologize, Mr. Shelby. I would not have interrupted, except that we have a bit of a situation on our hands. May I please borrow Ravi for a few minutes to help me sort it out? Otherwise, some folks are going to be in rather *serious trouble*," said Mr. Danzig, stressing his last phrase while he glared at Allen and Bill.

"Well, this is a bit irregular, but I suppose Ravi doesn't really need the review anyway. Go ahead, Mr. Danzing, if that is alright with Ravi," replied Mr. Shelby.

Ravi got up from his desk and put his history notebook in his backpack, saying as he walked out of the classroom, "Thank you Mr. Shelby. I'm sorry about this."

A bit calmer now, Mr. Danzing started to explain the situation to Ravi:

"Sorry about that Ravi. I was about to escort Mr. Kaitelle and Mr. Hennings here to the principal's office for cheating, when they pleaded with me to get your input and that you would be able to clear them."

"What's this all about?" asked Ravi, visibly concerned and surprised. Allen and Bill were both close friends of Ravi and his teammates on the Mathletes team. Ravi knew that they were both excellent math students who would neither need, nor wish, to cheat. Allen nervously began to speak, "Ravi, we tried to explain to Mr. Danzing that we had a solu..." but he was sharply interrupted by Mr. Danzing.

"Here's the problem, Ravi," began Mr. Danzing. "Last month, I told the students in my precalculus class that if they came up with and solved a truly innovative math problem, they would not have to take the final exam, and their average for the semester would count as their grade. Apparently, Mr. Kaitelle and Mr. Hennings have attempted to take advantage of my kindness by cheating!"

"How did they cheat, Mr. Danzing?" asked Ravi, hoping that there must have been some misunderstanding. Mr. Danzing began answering and was once again becoming increasingly agitated as he spoke.

"Mr. Kaitelle and Mr. Hennings told me they had a neat problem which they had designed and solved. I asked them to let me see it, but they stated that a demonstration would be more impressive. Allen came up to my desk and told me a positive integer, which he asked that I keep to myself. Bill then came up to my desk and told me another positive integer. They both claimed that they had chosen the numbers independently, and that neither of them knew what the other's number was. They then instructed me to write on the board, in no particular order, two numbers, one of which was supposed to be the sum of the two integers which they had told me and the other was supposed to be any number of my choosing, and that I was not to tell them which was which. I did as they asked. Bill then turned to Allen and asked, 'Do you know what my number is?' Allen replied that he did not, and then asked Bill, 'Do you know what my number is?' Bill replied that he did not. They went back and forth like this a few times, each of them repeating the same question, and replying that they did not know the other's number. Then suddenly, when Allen asked Bill, Bill said he knew Allen's number, and proceeded to tell me Allen's 'secret' number," concluded Mr. Danzing with a note of sarcasm, making quotation marks in the air with his fingers as he said the word "secret."

"How do you think they did that, Mr. Danzing?" asked Ravi, as if talking to himself.

Mr. Danzing replied, gesturing with his hands, "Obviously, they cheated! I think they had chosen the numbers together beforehand and then put on this little show for effect. But it's ridiculous—if you don't know the other person's number, how is repeating over and over that you don't know the number going to help you?"

"Would you mind if we go back to your classroom and try something, Mr. Danzing?" asked Ravi, trying to speak in the most soothing tone possible so as to help calm Mr. Danzing. Because of the tremendous respect that Mr. Danzing had for Ravi, he agreed. Everyone went back to the empty classroom, and Ravi asked Allen and Bill to sit at opposite ends of the classroom. Then, Ravi asked Mr. Danzing to choose any positive integer he wished and to write it on a piece of paper. Mr. Danzing chose 3,862. Ravi folded the paper and took it to Allen, saying "This is your secret number, Allen." Ravi had Mr. Danzing repeat the procedure, and Mr. Danzing wrote 4,139 on another piece of paper, which Ravi folded and walked over to Bill, with the same instruction as before.

Ravi then asked Mr. Danzing to write two numbers on the blackboard, in no particular order, one of which was the sum of the two numbers he had picked for Allen and Bill and the other being any number of his choosing. Mr. Danzing went to the board and wrote down 8,215 and 8,001. He then strode back to his desk, eyeing Allen and Bill, and sat in his chair with his arms folded across his chest.

Ravi then turned to Allen and asked, "Do you know what both numbers are, Allen?"

"No, I don't," replied Allen.

Then it was Bill's turn: "Bill, do you know what both numbers are?"

"No, I don't," answered Bill.

Back and forth it went for many turns, with Mr. Danzing becoming more convinced each time that since he had chosen the numbers himself, and the boys were unable to collude, the problem was unsolvable and that the boys were just stalling. Just as Mr. Danzing was about to get up to put an end to this charade, Ravi asked, "Bill, do you know what the numbers are?"

"Yes, I do!" exclaimed Bill, with visible relief, and proceeded to reveal not only his own number but Allen's as well.

Mr. Danzing shot up from his desk, saying "Let me see those papers," referring to the numbers he had written down. Both boys turned in the papers, and Bill was indeed correct that he knew both numbers. "How's it done?" asked Mr. Danzing. "What's the trick? Is it your voice? Like '*No*, I don't know the numbers,' means thousands, but 'No, *I* don't know the numbers,' means hundreds? Is that it? It's done with syllables, isn't it?" Mr. Danzing, by now, had worked himself into a frenzy.

"I could hear it in your voices, you know. I just haven't figured it out yet, that's all," he nearly shrieked, wagging his index finger at Allen and Bill, who by now looked rather horrified.

"Mr. Danzing ... Mr. Danzing," repeated Ravi calmly but emphatically, until he got Mr. Danzing's attention. Mr. Danzing stopped gesturing and looked at Ravi, a bit embarrassed, realizing that he had gotten carried away yet again.

"There is no trick, Mr. Danzing. It's legitimate mathematics. Allen and Bill have indeed come up with a beautiful problem. Please give me just a few minutes to convince you," said Ravi as he walked over to the blackboard.

Ravi then showed Mr. Danzing the mathematics behind Allen and Bill's "trick." Can you do the same?

Analysis and Hint

Let us analyze the problem at hand. The beauty of mathematics comes through most when it reveals the non-obvious, overturning what we believe by common sense to be clearly true and showing us the subtlety and elegance which lies beneath. Such is the case with this problem.

For ease, let us now refer to Allen as A and to Bill as B, and state the problem formally:

> Two boys, A and B, are each given a positive integer by their teacher, but neither boy knows the other's number. The teacher then writes two positive integers on the board and tells the boys that one of these two numbers is the sum of the boys' numbers, while the other is a number chosen at random. The teacher then asks A if he knows B's number. If A does not know, the teacher asks B the same question, and so on until one of the boys can tell the teacher the other's number.

Hint: The problem seems opaque, for how does A or B get any added information to figure out the numbers? The crux of the matter is to think hard about this query: Is there any extra information to be had when one of the boys answers that he does not know the other's number?

Solution

Assume that A and B are given different positive integers a and b, respectively, and that the teacher writes down the two numbers M and N, one of which is the sum of a and b. Let us say that $N < M$ and that the difference between M and N is d. Since we specify that a and b are positive integers, we know that $0 < a$ and $0 < b$.

The key to the problem is to realize that critical information gradually accumulates in the system with each negative answer and that each boy must keep track of this progressive accumulation.

Let us denote by A_k the conclusion that A could draw after the kth consecutive "No" of B. Similarly, B_k is the conclusion that B could draw after the kth consecutive "No" of A.

The first question is asked, and A answers "No." This gives B some information because A could have answered "Yes." If the lesser number on the board is less than or equal to a, A would have known that the sum of the two numbers could not possibly be N and therefore that the sum would have to be M (in other words, $b = M - a$ if $a > N$). Thus, we have

$$B_1 : a < N.$$

If B then answers "No," A can deduce some information about b. First, by the same chain of reasoning that established B_1, A deduces that $b < N$. However, B's "No," coming after A's "No," establishes an additional critical fact: $b > d$. Had B's number been lesser than or equal to d, B would know that $a + b < M$ by the following reasoning:

$$N + d = M \quad \text{(by the definition of } d\text{)},$$

$$a < N \quad \text{(fact } B_1\text{, known by both boys after A's first "No"),}$$

$$b \le d \quad \text{(by assumption).}$$

If these statements are true, we can sum the inequalities to get $a + b < N + d$. This is the same as $a + b < M$, and so $a + b$ would have to equal N.

Because B does not figure out the sum, A knows that b must be greater than d, and we have

$$A_1 : d < b < N.$$

This is a fact now known by both boys.

If A's next answer is "No," B can infer that $a < N - d$. If $a \ge M - d$, A would know that $a + b > N$:

$$b > d \quad \text{(fact } A_1\text{, known by both boys),}$$

$$a \ge N - d \quad \text{(by assumption).}$$

Summing again, we get

$$a + b > (N - d) + d = N.$$

A would then know that the sum must be M. Because A does not know the sum, B knows that $a < N - d$. Thus, we have

$$B_2 : a < N - d.$$

If B then answers "No," A can deduce that $2d < b < N$. If $b \le 2d$, B would know that $a + b < M$:

$$(N - d) + 2d = N,$$

$$a < N - d \quad \text{(fact } B_2\text{, known by both boys),}$$

$$b < 2d \quad \text{(by assumption).}$$

Once again, adding the left and right sides of the inequalities would give

$$a + b < (N - d) + 2d = N + d = M.$$

B would then have known that the sum of the two numbers must be N (because it cannot be M). Because B says "No," A knows that $b > 2d$, and we have

$$A_2 : 2d < b < N.$$

If A's next answer is "No," B can now infer that $a < N - 2d$. If $a \geq M - 2d$, then A would figured out which number was the sum:

$$b > 2d \qquad \text{(fact } A_2, \text{ known by both boys)},$$
$$a \geq N - 2d \qquad \text{(by assumption)}.$$

Adding the left and right sides of the last two inequalities would give

$$a + b > (N - 2d) + 2d = N.$$

A would then know that the sum must be M. Because A does not know this, B knows that the opposite of the above assumption must be true:

$$B_3 : a < N - 2d.$$

We can now see that a pattern has started to develop. With every consecutive "No," each of the boys is able to infer new information about the other's number and get closer to figuring out the sum. B's deductions follow the following pattern:

$$B_k : a < N - (k - 1)d.$$

A's deductions follow the pattern:

$$A_k : kd < b < N.$$

(These generalizations can be proved by induction (recall the definition of induction from "A Mystery on Sycamore Lane"), but we will leave that as an exercise for the reader.) Thus, we see that the bounds on a and b keep tightening with each consecutive "No." For example, A realizes that there are only two possibilities for b: $N - a$ or $M - a$. A then waits for his deductions to eliminate one of the possibilities, leaving him with the correct b. All the while, B is using the exact same method to deduce a.

Eventually, one of the boys will answer "Yes," and the process will terminate. With each round of negative responses, k increases (at a constant rate). Since we are dealing with finite positive integers, we must reach a point when one of the following happens.

1. The product kd reaches or exceeds b. Then, B would conclude that $a + b = N$ and terminate the game.
2. The product $(k - 1)d$ reaches or exceeds $N - a$. Then, A would conclude that $a + b = N + d = M$ and terminate the game.

This process will make significantly more sense when we look at the example given in our problem. In the story, $a = 3862$ and $b = 4139$. The teacher wrote 8001 and 8215 on the board. The teacher (or Ravi) then begins asking each of the boys in turn if they know what both numbers are.

The two students start with the following information:

A knows	B knows
$a = 3862$	$b = 4139$
$a + b = 8001$ or $a + b = 8215$ or $b = 4139$ or $b = 4353$	$a + b = 8001$ or $a + b = 8215$ or $a = 3862$ or $a = 4076$
$d = 8215 - 8001 = 214$	

Since A's first answer is "No," B can easily figure out that $B_1 : a < 8001$. If $a \geq 8001$, A would know that the sum of the two numbers could not possibly be 8001 and therefore that the sum would have to be 8215, and he would have been able to deduce b by subtraction.

If B then answers "No," A can deduce that $A_1 : 214 < b < 8001$. Had B's number been lesser than or equal to 214, B would know that $a + b < 8215$ (since both boys now know that $a < 8001$). Thus, B would have figured out that the sum would have to be 8001 (and been able to figure out a by subtraction).

Because B does not figure out the sum, A knows that b must be greater than 214. B can reason as well as A, and so he knows that A knows this fact.

If A's next answer is "No," B can infer that $B_2 : a < 7787$ (which is $8001 - 214$). If a was greater than or equal to 7787, A would have made the following deduction:

$b > 214$ (fact A_1, known to A and B),

$a \geq 7787$ (by assumption),

$a + b > 7787 + 214 = 8001.$

Thus, A would have deduced that the sum must be 8215. He did not, so B now knows that $a < 7787$, and A knows that B knows this.

If B then answers "No," A can deduce that $A_2 : 428 < b < 8001$. If $b \leq 428$ (i.e., 2×214 or $2d$), B would have figured out the following:

$$a < 7787,$$
$$b \leq 428,$$
$$a + b < 7787 + 428 = 8215.$$

Because B does not deduce this, A now concludes that $b > 428$ (and B knows that A knows this).

If A's next answer is "No," B can infer that $B_3 : a < 7573$ (i.e., $8001 - (2 \times 214)$). If $a \geq 7573$, then A would know that

$$b > 428,$$
$$a \geq 7573,$$
$$a + b > 7573 + 428 = 8001,$$

and that the sum would therefore have to be 8215.

We can now see our familiar pattern develop. With every consecutive "No" from A, B's deductions develop along the following pattern:

$$B_k : a < 8001 - 214(k - 1).$$

Meanwhile, with every consecutive "No" from B, A's deductions follow the pattern

$$A_k : 214k < b < 8001.$$

The process ends when $214k \geq 4139$ and B answers "Yes" or when $214(k - 1) \geq 4139$ (i.e., $214k \geq 4353$) and A answers "Yes." In this case, B figures out a and is able to answer "Yes" before A can deduce b. We confirm this with the following calculations:

$214k \geq 4139$ $214k \geq 4353$

$k \geq \dfrac{4139}{214} = 19.3411...> 19$ $k \geq \dfrac{4353}{214} = 20.3411...> 20$

$214 \times 20 \geq 4139$ $214 \times 20 \leq 4353$

Therefore, on B's 20th deduction (after A's 20th consecutive "No"), B learns that $a < 4066$. He reaches this conclusion as follows:

$$a = 3862 \text{ or } a = 4076,$$
$$B_{20} : a < 8001 - 214(20 - 1) = 4066,$$
$$a \neq 4076 > 4066.$$

With this last deduction, B finally knows that 4076 is above the limit for possible values of a and so a must be 3862.

Before we end this taxing journey, let us understand why A had not yet solved the problem. When $k = 19$, both A and B have said "No" 19 times and A is able to deduce

$$A_{19} : 214(19) < b < 8001, \text{ or } 4066 < b < 8001.$$

Now, A also knows that $a = 3862$. Thus, the two possibilities for b are $8001 - 3862 = 4139$ and $8215 - 3862 = 4353$. We see that both of these possibilities fall within the bounds which A has thus far deduced: $4066 < 4139 < 4353 < 8001$. Thus, A cannot yet distinguish which of the two possibilities is b.

The Urban Jungle

"It's okay, Ravi," said Ravi's mother, attempting to console her obviously disappointed son. For weeks, Ravi had been anticipating the opening of a new modern art exhibit at the Chicago Art Institute. As part of this exhibit, the famous artist David Melby was to unveil his new masterpiece of interactive art: *Urban Jungle*. Ravi had read an article in *Modern Art Digest* profiling Melby's passion for natural beauty and his distress regarding its ever-receding frontier in the face of urbanization. This exhibit was designed to underscore his views on the sterility and barrenness of modern cities and his disdain for their attempts to pretend that they retain some of nature's beauty by maintaining bits and pieces of parks and manicured lawns within their concrete confines.

The exhibit was strikingly simple in its design, yet supposedly also surprisingly compelling as one walked inside it. The best way to picture the exhibit was to visualize a large grid on the ground, like a big piece of graph paper, although no marks were actually on the ground. In the center, called ground zero, was a smooth circular piece of concrete, one foot in diameter. The exhibit itself was a large circular "forest," 200 feet in diameter, with ground zero at its center. This "forest" was made of modern "trees," each of which was a perfectly cylindrical smooth wooden stick 10 feet high and 1 inch in diameter. These sticks were spaced precisely two feet apart along the imaginary lattice points on the ground (see Figure 1). The two foot corridor between the rows and columns of the "forest" easily left room for a person to walk through and to contemplate the barrenness of the seemingly endless array of plain smooth sticks as compared to the appearance of a lush, green forest of real trees. The analogy between this urban jungle and a real forest was, in Melby's mind, precisely the analogy between man-made beauty and natural beauty.

Ravi's mother possessed a genuine appreciation for art, which she had successfully transmitted to her son. Although a strange concept for many, Ravi saw a great harmony between mathematics and art. Both required a keen sense

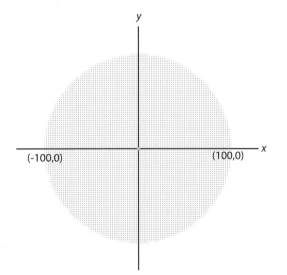

Figure 1. A bird's-eye-view diagram of *Urban Jungle*.

of aesthetics. Today, Ravi and his mother had planned to go to the opening of the *Urban Jungle* exhibit. However, early in the morning, Ravi's father, a Chicago District Attorney, called them from his office telling them that the exhibit was closed because of a police investigation at the site.

Later that night at dinner, Ravi's father told his family the shocking details of the day's events. Early in the morning, before the exhibit was set to open, there had been a shooting at the eastern boundary of the *Urban Jungle*. Steven Jennings, a noted art critic for the *Chicago Herald* who had been previewing the exhibit, was shot dead. The police now had Melby in custody for the shooting, which was witnessed by Simon Sullivan, another artist whose piece, *Bells and Whistles*, was also a part of the exhibit set to open that morning.

"That's very disappointing, Dad," said Ravi. "That doesn't sound at all like the Melby whose interview I read in *Modern Art Digest*."

"Well, Ravi, people will fool you. If there's one thing I've learned in this business is that most people cannot be trusted. Their image is one thing and their true self is another," replied Ravi's father.

"Still, I find it hard to believe," replied Ravi sadly.

"That's the good thing about hard evidence, Ravi. It takes away the element of subjectivity. We have a credible witness who saw Melby pull the trigger," said Ravi's father, sounding every bit the district attorney.

"Still, Dad, what about motive? Why would Melby kill Jennings?" inquired Ravi.

"According to Sullivan, Jennings was about to write a very scathing review for the *Chicago Herald* regarding Melby's exhibit, calling the *Urban Jungle* a rudimentary work without artistic merit. That's certainly motive enough," replied the father.

"Did Melby confess to the killing?" asked Ravi.

"No, Ravi, they rarely do. He claimed he was on the other side of the *Urban Jungle* when he heard a shot. He says he then ran around the exhibit and found Jennings in a pool of blood with a gun next to him. Melby claims that, without thinking but following a reflex, he picked up the gun. Several people arrived at the scene at this point and saw him standing over Jennings with a gun in his hand," answered Ravi's father.

"Any other witnesses who saw him pull the trigger, Dad? Surely, there must have been other people around the exhibit," asked Ravi.

"No, Ravi. This was about 6:40 a.m. Very few people were around. Those who heard the shot came to the scene and saw Melby standing over the body like I told you, but no one other than Sullivan saw him pull the trigger," explained his father.

Ravi was quiet for a bit, looking off into the corner of the room. Finally, he got up to go to his room. As he was about to leave the dining room, he turned to his father and asked, "Dad, where was Sullivan when he saw all of this? Did other people see Sullivan around the exhibit?"

His father answered, "No, Ravi, no one else saw him. They were, of course, distracted by the sight of Jennings *dying* in front of them. Anyway, Sullivan told the police that he had gone into the *Urban Jungle* exhibit to experience it for himself and was standing at 'ground zero' at the center of the exhibit. He says he heard a loud argument. Turning in the direction of the raised voices, he saw Melby pull out a gun and shoot Jennings."

"Thanks, Dad," said Ravi as he walked out.

"That Ravi! Sometimes he just has too much faith in people," Ravi's father said, turning to Ravi's mother and reaching for a slice of pie. Before he had transferred the slice to his plate, Ravi returned to the dining room.

"Dad, don't charge Melby. Sullivan is a liar, and I can prove it!" exclaimed Ravi.

Analysis

The question before us is: How did Ravi know that Sullivan was a liar? Naturally, it is because Sullivan made a claim that was mathematically impossible—that from the center of the *Urban Jungle*, he was able to see out to the rim. Assume that there is a circular "forest" of radius 50 units, constructed like the Urban Jungle, with "trees" planted at each of the lattice points except at the origin. These "trees," once again, consist of thin uniform vertical cylinders of a given radius r, spaced one unit apart. It turns out that if the radius of these cylinders exceeds 1/50 of a unit, then someone standing in the center of the circular forest cannot see outside of it no matter in which direction they look. Can you prove that, as Ravi did?

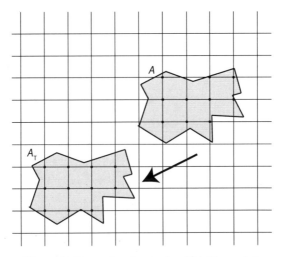

Figure 2. The region A covering 10 lattice points and translated to cover 11 lattice points.

Solution

This problem has probably the most intricate solution of all of the problems in this book (yet, paradoxically, the solution involves essentially no mathematical equations). The solution is intricate not because the mathematics is harder, but because it is deeper. What does this mean? It means that the solution takes multiple steps to unfold, each step of which builds directly on the one before. With such solutions, it requires some effort to follow each step, and more effort to follow the connection of one step to the next.

To begin the journey, let us prove a theorem in geometry called Blichfeldt's Lemma. Suppose that we have a piece of graph paper where the squares demarcated by the grid lines have an area of one square unit. If on this paper we draw a (bounded) region A with an area greater than n square units, Blichfeldt's Lemma states that it is always possible to translate this region A (i.e., to move it vertically and horizontally without turning it) to a position on the graph paper such that it will contain $n + 1$ lattice points. For example, if we draw an arbitrary region A with an area of 10.2 square units it can be translated to cover 11 lattice points (see Figure 2).

We can prove Blichfeldt's Lemma by considering the following scenario. Assume that we paint our region A gray, while the rest of the graph paper is white. Now, let us cut the part of the graph paper that contains the region A into unit squares along the vertical and horizontal grid lines. This will produce some number m of unit squares, where we know that $m > n$, since the area of A is greater than n. In our example, some of these unit squares will be completely gray while others will be only partly gray because they are

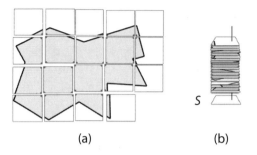

Figure 3. (a) Region A cut along the grid lines.
(b) Those unit squares stacked over S.

not entirely contained within the region. (See Figure 3(a).)
Having done this, let us imagine stacking these squares on
top of each other on a distant square S without changing
their orientation. This means that the bottom left corner,
for example, of each of our cut-up squares coincides with
the bottom left corner of S. This can be achieved by trans-
lating (sliding vertically and horizontally without turning)
each of our cut-up squares until it lies over S. Now, square
S has m unit squares stacked on top of it, and the gray
areas on those squares add up to the area of region A.

Next, consider the base square S itself onto which we
moved all the m unit squares. Imagine that a vertical ray
originates from each of the infinite number of points in
square S and pierces the corresponding point of each of the
m layers stacked on top of S. As the ray pierces each layer,
it either goes through a white point or a gray point. (See
Figure 3(b).) Suppose that no such ray passes through more
than n points. If this was the case, then each point in the
base S corresponds to at most n gray points, meaning that
the area of A could not be greater than n. Therefore, one
(or more) of these rays must go through at least $n + 1$ gray

points; let us call this ray r. Let us now drive an infinitely thin pin directly through each of the m unit squares in the location of ray r. This marks a point in each square m that has the same relative location to the lower left-hand lattice point of each unit square.

Now, let us unstack the m squares and reassemble region A. Each square now has a pinhole in the same relative location to the lower left-hand corner of that square. We know that at least $n + 1$ of these pinholes go through a gray point. All we need to do is to translate A until one of these pinholes covers a lattice point. Then, because the pinholes are in the same relative location in each unit square, each pinhole now lies on a lattice point. Because at least $n + 1$ pinholes are through gray points, the figure has now been translated to cover $n + 1$ lattice points, and Blichfeldt's Lemma is proven.

Now, let us derive a corollary of Blichfeldt's Lemma for the case $n = 1$:

> For a region A whose area is greater than 1, there are two points, P_1 and P_2, inside the region whose run and rise are both integers. In other words, if the coordinates of P_1 are (x_1, y_1) and the coordinates of P_2 are (x_2, y_2), then $(x_2 - x_1)$ and $(y_2 - y_1)$ are both integers.

This is easy to prove with the aid of Blichfeldt's Lemma. If the area of region A is greater than 1, then we know that it can be translated to cover at least two lattice points (see Figure 4). Let us perform the necessary translation and call the two lattice points P_1 and P_2 as above. Because they are lattice points, we know that the values x_1, y_1, x_2, and y_2 are integers and thus that $(x_2 - x_1)$ and $(y_2 - y_1)$ are integers. We then translate the two points using the reverse of the original translation so that the points now sit inside the original

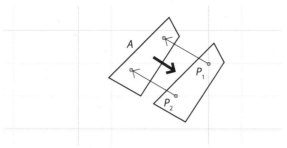

Figure 4. A visual representation of the proof of the corollary.

region A. Because we used the same translation for both points, the rise and run between them has not changed. Thus, we have identified two points in A whose run and rise are both integers.

All of this work has been building up toward a proof of a result known as Minkowski's Theorem, which states the following:

> A plane, convex region symmetric about the origin with an area greater than 4 square units covers a lattice point besides the origin.

In the above theorem, the word *convex* simply means that if we join any two points in the region by a line segment, that line segment is always inside the region. Circles and squares, for example, are convex regions. The term "symmetric about the origin" implies what we would intuitively think it does. To give it a formal definition, we would say that if a region contains point P with coordinates (x,y), it also contains point P' with coordinates $(-x,-y)$. Thus, what Minkowski's Theorem actually implies is that a convex plane region symmetric about the origin with an area greater than 4 will contain *two* lattice points, one which we label P_1 with coordinates (x_1, y_1)—where x_1 and y_1 are integers because this is a lattice point—and its mirror image lattice point P_1' with coordinates $(-x_1, -y_1)$. This is true because the region is symmetric around the origin.

To prove Minkowski's Theorem, let us begin with our convex region A with area greater than 4, symmetric about the origin (see Figure 5). Now, let us shrink this region by one-half in all dimensions. This is known in math as a *dilatation* with a *factor* of 1/2. Once again, this process is as intuitive as it sounds. We take each point of the region and bring it in along the line connecting this point with the origin until it is only half as far from the origin as it used to be. In more technical terms, for each point (x, y), we translate it to point $(x/2, y/2)$. This process precisely preserves the

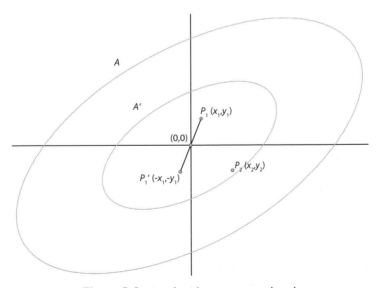

Figure 5. Region A with area greater than 4, and A shrunk by 1/2 in all directions.

shape of the region but makes it only half as big in terms of linear dimensions, equivalent to putting region A in a copy machine and printing a copy of it at 50% of its size.

Since we shrunk the region by one-half in terms of linear dimensions, we know that its area is now 1/4 of what it used to be. Since region A had an area greater than 4, the new shrunken region, which we call A', has an area greater than 1. By the corollary to Blichfeldt's Lemma, we know that it contains two points P_1 (x_1, y_1) and P_2 (x_2, y_2) whose rise $(x_2 - x_1)$ and run $(y_2 - y_1)$ are integers. Now, since A' contains point P_1, it also contains point P_1' $(-x_1, -y_1)$ because the region is symmetric about the origin. If we draw a line connection P_1' to P_2, this line is part of A', because A' is convex.

Let us now look at the midpoint of the line connecting P_1' and P_2 (see Figure 6). Recall that if we have two points in the plane, the midpoint of the line segment connecting those two points has x and y coordinates which are the average of the x coordinates and y coordinates of the two points. Therefore, the midpoint of P_1' and P_2 is point M with coordinates

$$\left(\frac{x_2 - x_1}{2}, \frac{y_2 - y_1}{2} \right).$$

Since A' is convex, the line between P_1' and P_2 lies in A'; thus, M is also in A'.

Now, let us re-expand A', dilating it by a factor of 2, to return it to the original region A. Point M is therefore sent to a new point N, whose coordinates are $(x_2 - x_1, y_2 - y_1)$. We already know that these coordinates are integers by the corollary to Blichfeldt's Lemma—they represent the rise and run between our two special points P_1 and P_2. Therefore,

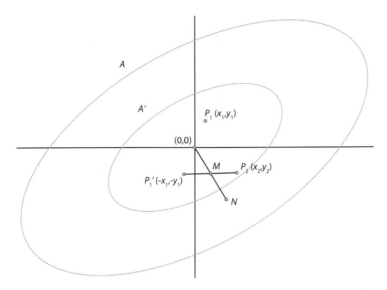

Figure 6. The midpoint M of P_1' and P_2 and $N = 2 \times M$, both in region A.

point N is a lattice point contained in A. Because A is symmetric to the origin, lattice point N, the mirror image of point N with respect to the origin, is also contained within region A. Thus, we have proven Minkowski's theorem.

Now, we are ready to solve our original problem. Before we do, though, let us briefly summarize the conceptual steps we have taken to get to this point:

1. We proved Blichfeldt's Lemma, that a plane region A which has an area in excess of n units can be translated to cover $n + 1$ lattice points.

2. Using the value of $n = 1$, we derived a corollary of Blichfeldt's Lemma, that for a plane region A with area greater than 1, there are two points within this region $P_1\,(x_1, y_1)$ and $P_2\,(x_2, y_2)$ such that $(x_2 - x_1)$ and $(y_2 - y_1)$ are both integers.

3. Using this corollary, we proved Minkowski's theorem
    that a convex plane region symmetric about the origin
    with area greater than 4 will contain a lattice point N
    other than the origin. Because the region is symmetric
    about the origin, it will also contain the mirror image
    lattice point N'.

At this point, we are ready to tackle the problem of the
Urban Jungle. We want to show that if the radius of the
trees, located at lattice points which are one unit apart, is
greater than 1/50 of a unit, then it is impossible to see out
of the circular jungle (with a diameter of 100 units) if we
are at the origin.

Let us start at the origin and look in the direction of
some arbitrary point P on the circumference of the circular jungle, which has radius R. (See Figure 7.) Let us label
the mirror image point of P with respect to the origin as
point Q. (Note that Q is also on the circumference.) The line

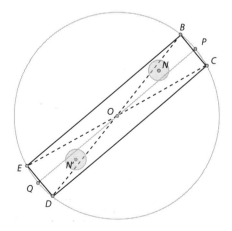

Figure 7. Trying to look out of the jungle from the origin.

PQ, which passes through the origin O, represents an arbitrary diameter of the circular jungle. Now we draw a chord BC such that the length of BC is 4(1/50) = 2/25 and BC is bisected by PQ. (Thus, the distance from point B to the line PQ is 1/25, and the same is true for point C.) Let us take the mirror images of B and C with respect to the origin and call them D and E, respectively. Note that the length of DE is also 2/25, and, because we are dealing with mirror images around the origin, BC and DE are parallel.

The four points create a rectangle $BCDE$ that is completely contained within the circle. We know that $BCDE$ is a rectangle because the diagonals BD and CE have equal length (they are both diameters of the circle) and they bisect each other (they cross at their mutual midpoint, the origin). Therefore, the angle BCD is a right angle, and we can calculate the length of CD—let us call it l—by using the Pythagorean Theorem:

$$(2R)^2 = (2/25)^2 + l^2,$$
$$l = \sqrt{(2R)^2 - (2/25)^2}.$$

The area of the rectangle $BCDE$ is

$$(2/25)l = (2/25)\sqrt{(2R)^2 - (2/25)^2}$$
$$= (2/25)\sqrt{4(R^2 - (1/25)^2)} = (4/25)\sqrt{R^2 - (1/25)^2}.$$

We know that $2R = 100$, and we can show that the area of the rectangle is greater than 4:

$$(2/25)l = (4/25)\sqrt{R^2 - (1/25)^2}$$
$$> (4/25)\sqrt{(R - 1/25)(R - 1/25)} = 4/25(R - 1/25)$$
$$> (4/25)(50 - 1/25) = 4(2 - (1/25)^2) > 4.$$

By Minkowski's theorem, $BCDE$ contains two mirror image lattice points, N and N'. A tree is centered at each of these points. Since the radius r of the tree is bigger than 1/50, the diameter of the tree is bigger than half the width of the rectangle $BCDE$. This means that the tree at N cuts off the line of sight from the origin to point P. Similarly, the tree at N' cuts off the line of sight in the direction of Q. Since P was an arbitrary point on the circumference of the circle, the same argument is valid for any point on the circumference, i.e., for any direction in which one looks from the origin. This proves our contention that it is impossible to see out of the circular jungle when standing at the origin.

In our particular problem, the jungle was 100 feet in radius, with trees spaced at lattice points 2 feet apart. Thus, we will take 2 feet = 1 unit. Therefore, our jungle has a radius of 50 units with trees spaced at lattice points 1 unit apart. Each of the trees is a cylinder 1 inch in diameter, or 1/2 inch in radius. Given that 1 unit = 2 feet = 24 inches, each tree thus has a radius of 1/48 of a unit, just slightly larger than 1/50. Therefore, Ravi realized that it would be impossible for Simon Sullivan to see out of the jungle while standing at "ground zero," directly contradicting the statement that Sullivan gave to the police. Ravi's proof prompted the police to closely analyze the contents of Jennings' computer files. Jennings' notes showed that he was, in fact, getting ready to write a terrible review of Sullivan's exhibit *Bells and Whistles*, and that he had nothing but praise for Melby's *Urban Jungle*.

Extension

It is interesting how often a case hinges on precise analysis. Sullivan said that, while at ground zero, he saw a murder committed along the eastern border of the jungle. The

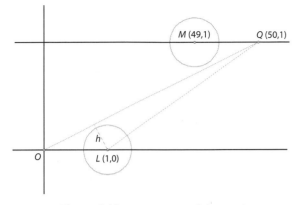

Figure 8. How to see out of the jungle.

trees, as it turned out, were just a touch too large to make his story plausible. As we saw above, if the trees were any larger than 0.96 inches in diameter (i.e., if their radius was any larger than 1/50 of 2 feet), it would be impossible to see out of the jungle. As it turned out, Melby's trees were precisely 1 inch in diameter.

Now, we would like to show that if the radius is ever so slightly less than 1/50 of a unit—$\frac{1}{\sqrt{2501}}$ to be precise—it is possible to see outside the jungle by looking in the right direction. That direction is along the line joining the origin to the lattice point Q (50, 1). This line of sight makes a very slight angle above the x-axis and would allow the viewer to see an event happening at the eastern boundary of the jungle, as Sullivan had claimed.

Let us look at Figure 8, which has been enlarged and exaggerated to help us visualize the situation. The length of line segment OQ is

$$\sqrt{2501} = \left(\sqrt{1^2 + 50^2} \right).$$

It is easy to see that the two lattice points closest to this line segment are L (1, 0) and M (49, 1), both of which are symmetrically located with respect to line segment OQ. If the trees located at these lattice points do not block the line of sight along OQ, no other tree will. The points $O, L,$ and Q form the triangle $\triangle OLQ$. We will calculate the area of this triangle in two ways:

1. If we use OL as the base, then the area of $\triangle OLQ$ is $1/2 \times 1 \times 1 = 1/2$.

2. Now, let us use OQ as the base and the perpendicular from L to OQ as the height h. Thus, we get that the area of $\triangle OLQ = 1/2 \times OQ \times h = 1/2$. Recalling that $OQ = \sqrt{2501}$, we get

$$\frac{1}{h}\sqrt{2501} = \frac{1}{2}, \text{ or } h = \frac{1}{\sqrt{2501}}.$$

Thus, if a tree has radius $r < \frac{1}{\sqrt{2501}}$, it is too small to intersect the line of sight OQ and would not block the view along that line. Given that each of our units was 2 feet = 24 inches, this upper limit for r is $\frac{1}{\sqrt{2501}} \times 24$. This is equivalent to a diameter of 0.9598 inches. Thus, if the diameter of the trees had been anything less than this, say 0.95 inches instead of 1.00 inch, Sullivan's story to the police would have been entirely plausible and could not have been refuted by Ravi. Hence, the solution to this case hinged on precision.

A Snowy Morning on Oak Street

A thick, white blanket of snow covered Oak Street in Chicago's suburbia, where Ravi lived with his family. Even by Chicago's standards, this amount of snowfall was remarkable. Sometime before dawn, it was as if the sky had just opened up and started dumping snow. The rate of snowfall was steady at what meteorologists call MPR (maximal precipitation rate), and there was no let-up in sight.

Ravi walked to the edge of the driveway and looked out on the street, amazed by the snowfall. He could barely see a few feet in front of him from the haze of the swirling flakes, falling like chunks of cotton candy. He was dressed in his coat and his snow cap was pulled down over his ears. His bookbag was slung over his shoulder. It was 8:00 a.m., and he was waiting for his father to drive him to school; there would be no walking today. As Ravi stood at the edge of the driveway, he heard a loud clanking, and out of the haze of snowflakes, a large snowplow gradually appeared, slowly lumbering along Oak Street, clearing snow to the edge of the road. When Bert McGillycuddy saw Ravi's driveway, he halted and looked over, putting a gloved hand to his eyebrow and yelling out, "Hey Ravi, is that you?" Although Ravi was only a few feet away, visibility was so poor that Mr. McGillycuddy was not sure it was him.

"Hello, Mr. McGillycuddy," yelled Ravi over the din of the snowplow's roaring engine. Ravi was an affable young man with a pleasant and easy-going manner. Although he was extremely intelligent and from a relatively wealthy family, he easily made friends with people from all walks of life. He counted among his friends bus drivers, mail carriers, butchers, museum guards, and, in the case of Mr. McGillycuddy, snowplow operators.

"How do you like this weather?" yelled Mr. McGillycuddy.

"It's unbelievable! I've never seen anything like it," answered Ravi.

"Yeah, I can't believe how fast it's coming down," said Mr. McGillycuddy. "I've been working steadily for the last two hours, plowing snow as fast as this plow will let me. Between 6:00 and 7:00 a.m., I covered four blocks from River Road to Edgeview Avenue, but in the last hour, I've only been able to cover two blocks, from Edgeview to here," continued Mr. McGillycuddy.

Just as Ravi was about to answer, he heard the beep of a car horn. He turned to see that his father had backed out of the garage and was waiting for him to get in. Ravi yelled out to Mr. McGillycuddy, "Sorry, Mr. McGillycuddy, my dad is waiting for me. I'll talk to you later. Good luck today." Ravi turned and ran to the car.

Later that night, at dinner, the family ate while watching the news. The airport had closed down at midday due to the snowfall, and the Chicago Unified School District had announced that tomorrow would be a "snow day"—the schools would close down because of the excessive snowfall.

"I guess everything comes to a halt with this sort of weather," said Ravi's mother.

"Everything except crime," replied Ravi's father, a Chicago district attorney.

"Why, what happened today?" asked Ravi, who took a keen interest in his father's hard-to-solve cases.

"Taper's Jewelry Store downtown was robbed. The safe was blown using drilled-in charges that were placed surrounding the door of the safe on all sides, creating sufficient pressure to blow the door open. Very sophisticated," said Ravi's father.

"Yes, Dad. It requires just the right distribution of charges to be able to make something like that happen. I imagine not many people know how to do something like that," replied Ravi, now finding the case even more interesting.

"You're absolutely right, Ravi. In this county, I figure that only Jimmy 'Pickles' Graziano and Hyun-Chun Kim would be able to do that, and Kim is already in prison," replied Ravi's father.

"So what's the problem, Dad? Why don't you pick up Pickles?" asked Ravi.

"It's not that simple, Ravi. I am very suspicious of him, as is Judge Sebastian. However, we don't have enough for a search warrant, let alone an arrest. All we were able to do is question him. The safe's alarm system at Tapers went off at 4:53 a.m. Pickles has an alibi starting at 5:30 a.m.—he was seen outside his apartment by old Mrs. Fisher, who had to go out and walk her dog. She says that even through the heavy snow, she recognized him under the street lamp at the entrance of the apartment complex," answered Ravi's father.

"How far is Graziano's apartment from Taper's Jewelry Store, Dad?" asked Ravi.

"We've already checked on that, Ravi. With no traffic, it might be possible to make the drive in about 25 minutes. But in this snow, with no visibility, it would have been absolutely impossible for Pickles to make it from Taper's back to his apartment by 5:30 if he were the perpetrator," explained Ravi's father.

"Just how much do you need for a search warrant, Dad?" asked Ravi.

"All Judge Sebastian needs is a reason, Ravi. Just a possibility that Pickles could be the perpetrator. We're sure he's our man," said Ravi's father, shaking his head with regret that a guilty man will have beaten the system.

"Remember this morning, Dad? You honked at me when I was talking to Mr. McGillycuddy. Based on what he told me, I think I have enough to get you that search warrant!" exclaimed Ravi, with a broad smile.

What did Ravi mean? From the details we have been given, how could he find a possibility that Pickles was guilty?

Hint: From the information given, Ravi was able to figure out what time it started snowing and that Pickles would have just had time to empty the safe and make it almost all the way home before the snow started falling. If you make a couple of reasonable assumptions, can you do the same?

Analysis

From the details we have, we can actually put together the elements of a very nice math problem, which we can state formally as follows:

> Sometime before 6:00 a.m., snow starts falling at a heavy, uniform rate. At 6:00 a.m., a snowplow begins to clear the street. The plow removes a fixed volume of snow per unit time. In the first hour, the snowplow moves a certain distance (in this case, four blocks, according to Mr. McGillycuddy). In the next hour, it moves only half the distance. What time did it start snowing?

Solution

Since the snowplow can clear a fixed volume of snow per
unit time, then, for a street of given width, the linear velocity of the snowplow up the street is inversely proportional
to the depth of snow on the ground. Now, we will employ
a clever use of variables that relies on the fact that the
unknown proportionality constant which determines the
snowplow's speed will end up canceling out.

We assume that it started to snow at some point $t = 0$.
We know that the rate of the snowfall is uniform (i.e., constant), so let's call that rate s and let's measure s in units of
depth added per time. Therefore, the depth of the snow at
time t is st units. The speed of the snowplow at time t will
then be proportional to $1/st$, say $k(1/st)$. This expression is
the same as $(k/s)(1/t)$. Now, let us say that the snowplow
started clearing Oak Street at some time $t = x$.

We know that, in general, $d = vt$, where d is distance,
v is velocity or speed, and t is time. Therefore, the distance
covered by the snowplow in some time interval t is vt. However, this equation is valid only under the assumption that
the velocity v is constant throughout the time interval t.
In our case, this assumption is clearly untrue since it is
snowing quite heavily, with the depth of snow gradually
increasing (and the snowplow's velocity gradually decreas-
ing) with time. Calculus was invented to deal with just this
sort of circumstance. In this case, we calculate the distance
d covered by the snowplow using the integral of the equation $v(t)$, which gives us the velocity of the snowplow at any
given time t, over the range of time starting at x and ending
at some time T:

$$d = \int_{x}^{T} v(t)dt.$$

If you are unfamiliar with integrals, the expression above basically takes the sum of the distance calculations $d = v(t)dt$, where dt is a small change in time, for $t = x$, $x + dt$, $xi + 2dt$, ... , $T - dt$, T.

Using the notation we've developed, we can say that the velocity equation is $v(t) = k/st = (k/s)(1/t)$. To calculate the distance covered in the first hour, d_1, we take the integral of $v(t)$ from $t = x$ to $t = x + 1$ (time is being measured in hours):

$$d_1 = \int_{x}^{x+1} \frac{k}{s}\left(\frac{1}{t}\right) dt.$$

Since k and s are both constants, we can pull k/s out of the integral and restate the equation as

$$d_1 = \frac{k}{s} \int_{x}^{x+1} \left(\frac{1}{t}\right) dt.$$

Now, we recall from calculus that the antiderivative of $1/t$ is $\ln(t)$. Therefore,

$$d_1 = \frac{k}{s}\ln(t)\Big|_{x}^{x+1} = \frac{k}{s}(\ln(x+1) - \ln(x)).$$

By precisely the same analysis, the distance covered in the second hour is

$$d_2 = \frac{k}{s}\ln(t)\Big|_{x+1}^{x+2} = \frac{k}{s}(\ln(x+2) - \ln(x+1)).$$

We know from Mr. McGillycuddy that the snowplow covered twice the distance in the first hour as it did the second hour. Thus, we know that $d_1 = 2d_2$ or

$$k/s(\ln(x + 1) - \ln(x)) = 2\ k/s(\ln(x + 2) - \ln(x + 1)).$$

We can cancel out the constant that appears on both sides of the equation and then simplify the expression as follows:

$$\ln(x + 1) - \ln(x) = 2(\ln(x + 2) - \ln(x + 1))$$
$$= 2\ln(x + 2) - 2\ln(x + 1),$$
$$3\ln(x + 1) = 2\ln(x + 2) + \ln(x).$$

Recalling that $\ln(ab) = \ln(a) + \ln(b)$ and that $a\ln(b) = \ln(b^a)$, we can rewrite the above equation as

$$\ln((x + 1)^3) = \ln((x + 2)^2 x)$$

Now, exponentiating both sides, we get

$$(x+1)^3 = x(x+2)^2 .$$

Now, we expand both sides to get

$$x^3 + 3x^2 + 3x + 1 = x^3 + 4x^2 + 4x .$$

Simplifying, we get the quadratic equation

$$x^2 + x - 1 = 0 ,$$

which has solutions

$$x = \frac{-1 \pm \sqrt{5}}{2} .$$

Only the positive solution makes physical sense, and so we get that

$$x = \frac{-1 + \sqrt{5}}{2} .$$

This represents the time interval in hours between the time that it started snowing and the time that the snowplow started its course. This works out to about 37 minutes, meaning that it started to snow at about 5:23 a.m.

This would have left Pickles just enough time to empty the safe and make it home before the snow started. This analysis was all that Judge Sebastian needed to issue a search warrant. The stolen jewels were recovered from Pickles' apartment, and he was arrested.

Before leaving this problem, we should stand for a while in wonder before the power of mathematics. We were able to figure out when it started snowing without knowing the depth of snow at any particular point in time, the rate at which the snowplow could clear snow, or the rate at which the snow was falling! This is completely amazing.

Conclusion

Well, we have come to the end of our journey. I sincerely hope that you enjoyed matching wits with Ravi. Doubtless, some of the stories may have seemed far-fetched, or even a bit contrived. However, my hope is that they made the mathematics behind them seem both enjoyable and practical. One of the most frequent phrases that I hear from friends exasperated by an evening's math homework is, "When am I ever going to use this stuff?" While they ask what they believe to be a rhetorical question, I hope that this book provides a very small glimpse of an answer. Once again, though, as I said in the preface, I do not like mathematics for its utility, but rather for its beauty.

I hope that, whatever you think of the stories, you found at least a few surprises which defied your initial impression or your intuition. If that is the case, then you have touched some of the fun and the magic that is mathematics. From here, the sky is the limit!

About the Problems

In this short appendix, I have tried to note where most of the problems on which the stories in this book were based can be found. This does not mean that the cited source is the *original* source of the problem. For most popular problems in mathematics, the original source is likely impossible to find. This index will also serve as a reading guide or reference section to what I think are some great books for those who are interested in mathematical problem solving.

A Mystery on Sycamore Lane

I have seen this problem several times. It appears with a nice discussion in the wonderful book *The Art and Craft of Problem Solving*, by Paul Zeitz (John Wiley and Sons, New York, 1999, pp. 84–85).

The Watermelon Swindle

I do not recall the source of this cute problem. I read it, and it just stuck with me.

An Adventure at the Grand Canyon

The problem here is based on a similar problem given in *The Art and Craft of Problem Solving* by Paul Zeitz (John Wiley and Sons, New York, 1999, p. 65). However, the problem is provided as an exercise, and no solution is given in that book.

Basketball Intrigue

There is no specific source for this problem, which I made up. Once again, it is a combinatorics problem, which is my favorite area of mathematics.

The Moon Rock

The source of this problem and most of the solution material is the fascinating book *Discovering Mathematics: The Art of Investigation*, by Anthony Gardiner (Oxford University Press, Oxford, UK, 1987, Chapters 12–15).

A Theft at Dubov Industries

The problem at the heart of this story is a very popular one that I have seen in a few places. It can be found with a nice solution in *Mathematical Chestnuts from Around the World*, by Ross Honsberger (Mathematical Association of America, Washington, DC, 2001, p. 31). The puzzle in the Extension comes from another one of Dr. Honsberger's many excellent books, *More Mathematical Morsels*, (Mathematical Association of America, Washington, DC, 1991, p. 235).

Murder at The Gambit

The problem here is based on a similar one found in the wonderful book *Math Charmers: Tantalizing Tidbits for the Mind* by Alfred Posamentier (Prometheus Books, Amherst, NY, 2003, p. 233). The material in Extension 1 is based, in part, on a solution by Stephen Kaczkowski to Problem 782 in *College Mathematics Journal*, Vol. 36, No. 4, September 2005, p. 334.

A Day at the Racetrack

There is no precise source for this problem—I made it up based on readings about combinatorics. The material on derangements can be found in any combinatorics book. I think one wonderful such book is *Mathematics of Choice* by Ivan Niven (Mathematical Association of America, Washington, DC, 1965). Extension 2 of this problem is pretty hard. The solution for the expected value portion is based upon a solution to a related problem that appeared in the 1987 International Mathematical Olympiad. The problem and solutions are found in *International Mathematical Olympiad 1959–1999*, by Istvan Reiman (Wimbledon Publishing, Wimbledon, UK, 2001, pp. 329–330).

Bowling Average

I have no specific source for this problem. It was posed to me for fun by Klaus Peters, and I have presented the solution with which I came up.

Caught on Film

The problem here is based on a similar problem from an old mathematics contest given in Hungary called the Eotvos Competition. That problem appeared on a test from 1900! It can be found in *Hungarian Problem Book I: Based on the Eotvos Competitions 1894–1905*, by J. Kurschak (Mathematical Association of America, Washington, DC, 1963).

A Mishap at Shankar Chemicals

The problem at the heart of this story came from a nice book of mathematical brainteasers called *The Unofficial IEEE*

Brainbuster Gamebook, by Donald R. Mack (IEEE Press, New York, 1992). Only a partial solution is given there.

Almost Expelled

This amazing problem can be found in the beautiful book *Mathematical Miniatures*, by Svetoslav Savchev and Titu Andreescu (Mathematical Association of America, Washington, DC, 2003, Problem 21).

The Urban Jungle

This wonderful problem, sometimes called the Orchard Problem, and its solution are found in Chapter 4 of *Mathematical Gems I*, by Ross Honsberger (Mathematical Association of America, Washington, DC, 1973, pp. 43–53) as well as in *Challenging Mathematical Problems with Elementary Solutions*, Vol. II, by A. M. Yaglom and I. M. Yaglom (Holden-Day Press, San Francisco, 1967).

A Snowy Morning on Oak Street

This problem has apparently been published in a couple of places. I saw it in *A Friendly Mathematics Competition*, edited by Rick Gillman (Mathematical Association of America, Washington, DC, 2003, pp. 96–97). However, the solution published there is a bit hard to follow, and the problems in the book are hard for high schoolers. The approach that I took is based on a clever solution posted by Michael Shackleford at http://mathproblems.info/prob2s.htm.

About the Author

Leith Hathout, an accomplished young mathematician and a mystery enthusiast, is a high school student in California. He began participating in national mathematics competitions at the age of 9, eventually winning several awards at the national level and achieving a perfect score on the California Math League exam. Outside of his studies, he excels at playing SET® and is a nationally ranked foil fencer.